WE'RE OVERDOSED

Barry I. Gold, Ph.D.

NTRAL PARK SOUTH PUBLISH

Published by Central Park South Publishing 2022
www.centralparksouthpublishing.com

Typesetting and e-book formatting services by Victor Marcos

ISBN:
978-1-956452-20-4 (pbk)
978-1-956452-21-1 (hbk)
978-1-956452-22-8 (ebk)

Contents

The Curse of Addiction

Let's Stop Cursing

My student, a blonde, young woman, perhaps thirtyish and not too tall but with sad eyes and no smile, put her tray down on my table, sat opposite from me, leaned toward me, and asked, "Why do you do teach this course?" She creased her forehead expectantly, emphasizing the question as she waited for my response. She was sincerely curious, and she demonstrated that curiosity by joining me at my table without my invitation.

In fact, my mouth was open as I began to bite into my sandwich. All I could do was nod and mutter, "Hi," as I chewed my mouthful.

She was one of my students in the weekend course I taught, *Pharmacology of Drugs of Abuse,* at Jersey City State College. I'll call her Sally. She took a sip from a cup of soup before continuing with, "I got clean two years ago." She took another sip and repeated in a voice that almost pled, "Really, why do you teach this course? Are you an addict too?"

The entire class was interested, and students followed her lead. They began pulling chairs from adjacent tables so they could join us. Before long, my entire class surrounded me. I hadn't invited anyone, but I was flattered because following me to the cafeteria and surrounding my table suggested they were enjoying my course and they were all curious about me.

Sally continued staring at me with those sad blue eyes as though waiting for my judgment. I suppose her eyes reminded me of my own daughter. I am not judgmental, so I simply continued to give her my attention as I chewed.

"Yeah," another student asked, "Are you an addict too?" It was clear they wondered about me, in part, they took turns explaining, because so many people had written them off in their lifetimes. They were all addicts in recovery, training to be drug counselors, and all had newly dedicated themselves to preventing others from heading down that dead-end street they all knew, Addiction Street. Their former, drug-using lifestyles made them all expect to receive attention only from other drug users or from law enforcement. Long ago, the rest of society had written them off as worthless junkies and they all carried that rejection as an open wound.

I swallowed and said, shaking my head, "No, I'm not an addict. I just like to teach." I explained with a smile and added, "I pay back community service and I do that because the government paid for my postdoctoral fellowship years before. This course also keeps me teaching young people, and I enjoy that too. I taught medical school when I began my career, and I was good at it, but teaching was the only part of being a faculty member I ever missed."

Paradoxically, as I think back, teaching was also the smallest part of my medical-school faculty membership earlier in my career. Back then, I recall lecturing four hours a year in our 2nd-year pharmacology course. I also mentored two graduate students, one for her Ph.D. during my six years at the university, so my lab was full. My one-day course for addiction-counselors-in-training was seven hours by itself, almost doubling in one day, my former teaching load for one year.

My medical students never voluntarily joined me for lunch either.

We were in the cafeteria at what is now *New Jersey City University* where I taught a weekend review of the pharmacology of frequently abused drugs to students working toward certification as addiction counselors. Teaching them was rewarding and fascinating, but still demanding because the course kept me working all weekend and added 2-hour commutes on Friday and Sunday. That schedule extended my workweek to seven days, and I also spent two weeks each month in Europe because I had responsibilities there. That part of my career meant I worked all day after flying all night.

I was younger then.

My livelihood came from managing two drug development programs, for central nervous system drugs and vaccines, we called them biologicals, in the U.S.A. and Europe. My position was at a large drug company whose headquarters was on what's commonly known as Philadelphia's Mainline. I enjoyed living on the Mainline and I owned a 200-year-old farmhouse on three acres that I occupied with my parents and my older son. I loved teaching, old houses and extended families, so I was a mostly happy man. My first marriage had ended a few years before I bought the farmhouse, and I was also dating a woman back in New Jersey, so I had a place to stay when I taught.

I lectured for three hours in the morning at Jersey City and broke for lunch, where it became a regular event that all my students joined me while I ate. I had been divorced five years by then, my first divorce of two, and I enjoyed the contact with younger people, in part because I had a few years of seeing my own kids less frequently because of that divorce. My son moved in with me, but I still lived 100 miles away from my daughter. My ex-wife Joan lived with her in New Jersey, and I lived in my Pennsylvania farmhouse with our son and my parents.

I was always struck that my addiction-counselor students displayed an intellectual interest in what I taught that was greater than the interest shown by the medical students I had taught earlier in my career when I was still on the faculty at the medical school in Bethesda, MD. The medical students were overwhelmed with their coursework and military training, while these addiction-counselors-in-training were overwhelmed by the choices they made that brought them to my course; their interest in my subject was palpable. Each trainee was starting over as a recovering addict, relishing the opportunity, fully aware of, but undaunted by the challenge of helping others and entirely self-motivated.

Their induced feeling of no or low self-worth, likely included family estrangement, peer drug use, and feeling out of place, have all been implicated as motivators of opioid drug use. "If you can't feel anything, nothing can bother you," was their common response if asked why they started using opioids.

They voiced that when anyone asked, "Why do you do drugs?" They could never articulate what feeling bothered them to the extent they had to block the feeling, preferring to feel nothing. "I was stressed and couldn't concentrate," was their common thread.

Dr. Gabor Maté, in his excellent book *In the Realm of Hungry Ghosts, Close Encounters With Addiction*, summarized the motivation to take opioids as, "Far more than a quest for pleasure, chronic substance use is the addicts attempt to escape distress." Even more succinctly, he wrote, "It's their attempt, I believe, to escape the hell realm of overwhelming fear, rage, and despair." Look deeply enough and opioid addicts could all describe a history of isolation, pain, or other psychological insult.

I would drive back to Philadelphia on Sunday night, exhausted, exhilarated, and hoarse. The people who took my course were there for varied reasons. Some were there because they were taught that to get clean, they had to give back. For others though, the course was more than their next step in their own recovery. Addiction counseling had become their calling.

I had asked Sally that day she sat down, "Were you sick and tired of being sick and tired?" She slowly nodded 'yes.' I had learned that question as I did my homework to prepare for the class. Then I asked her, "What made you start?" That has always been and still is the most interesting question to me whenever I prepare to teach, appear as an invited speaker, or interview a recovering addict.

But what is addiction? I'd thought about it the other day after I bought a two-and-a-half pound-jar (more than 1 kilogram) of salted cashews. Within a few days, I'd finished the jar. If I bought a bag of potato chips, I usually opened it and started crunching before I arrived home from the grocery store.

Am I addicted to cashews or potato chips?

The simple answer is no. They just taste good, feel good on my tongue and I like the salt and the crunch. Eating them, like any bad habit, is...just a bad habit.

If repeating an activity because it feels, tastes, or sounds good isn't an addiction, then what is?

The American Psychiatric Association defines addiction as "a brain disease that is manifested by compulsive substance use despite knowing there are harmful consequence."

The National Institute on Drug Abuse (NIDA), part of the U.S. National Institutes of Health, or NIH, defines it similarly, "as a chronic, relapsing disorder, characterized by compulsive drug seeking and use despite adverse consequences. It is considered a brain disorder..." That is a key concept, drug addiction is a disease, not a crime to be punished.

Those definitions have important ideas within them. Drug addiction is chronic and compulsive. There are harmful consequences and it's a brain disorder.

Stuffing myself with potato chips or salted cashews doesn't have harmful consequences other than adding to my weight, increasing my shopping bill, or making me thirsty, so I'm off the hook from addictive consumption. Craving can be reduced.

Opioid addicts have a sliding scale of harmful consequences and early in their addiction they might say, "I have to keep taking it because if I don't, I feel lousy."

Feeling lousy is the harmful consequence they try to avoid by taking the drug again and again. Although there are other harmful consequences, that theoretical monologue contains within it, part of the definition of addiction. It is substance abuse and the substances abused are drugs. Feeling lousy if they don't take them also hints at the other characteristic of opioid addiction; ceasing to take them results in withdrawal. That's the lousy feeling they try to avoid, the onset of withdrawal. It's unavoidable and the only way they know how to prevent it is to take the drug again.

NIDA, the National Institute on Drug Abuse (pronounced Nye-Da) says the main reason people take drugs is to feel good. Users say, "I enjoy getting high." Other users say, "I want to feel less stress." They also take drugs because they've heard a drug might improve their performance. Finally, people take drugs because friends, relatives, classmates, or roommates say, "Try this, it's great." I already mentioned Maté's comment, "...to escape the hell realm of overwhelming fear, rage, and despair."

The problem with all those reasons is that chronic use of some drugs causes addiction to them. Those drugs that produce addiction with chronic use are well-known and include the opiates morphine and codeine, heroin as well as the drugs nicotine, amphetamines, alcohol and some inhalants, cocaine, sedatives, hypnotics, and tranquilizers. The list is long but manageable and synthetic opioids deserve a spot at the top of the list next to the natural opiates morphine and codeine. The point is the drugs take hold of the user. Addiction sneaks up on them.

In the U.S., opioids are "scheduled," meaning they've been ranked according to their risk and assigned to a list of categories, developed by the *Drug Enforcement Administration* (DEA). The very top of the list, so-called Schedule 1 drugs, have no current medical use – legally, they cannot be prescribed by anyone - and they have a high potential for misuse. Heroin is a Schedule 1 drug along with LSD, Ecstasy, and marijuana, outlawed Federally since 1937. Many states, including New Jersey, my own state, have begun legalizing marijuana for in-state use only, including even recreational use. There are others too and all Schedule I drugs fit the description of having a high potential for misuse and no medical use. They are all in the news frequently. The surprise may be that alcohol is not on the same list as heroin and cocaine, although there are estimates that 14.4 million Americans have Alcohol Use Disorder. Alcohol is not scheduled in the U.S. because of the 21st Constitutional Amendment, ratified in 1933. It gave control of alcohol to the States and repealed the amendment that created and supported Prohibition. There are estimates that 60% of adult American men drink heavily or binge once a month. I don't have statistics for women drinkers.

Schedule 2 drugs are the next group on the list. Their entries are thought-provoking because the drugs not only have accepted medical use, that is, they can be prescribed by a physician, they also have a high potential for misuse. The list reads like the headline of a drug bust because it lists morphine, fentanyl, OxyContin, Demerol and Dilaudid, among others. Each of those drugs has medical purpose yet each is also traded in the illicit market because people abuse them routinely. One of the theories of OxyContin abuse, for example, was that its rampant abuse, in part, was fueled by its manufacturer, Purdue Pharma's, marketing efforts to woo physicians with speaking engagements and heavy reimbursement, as well as paid attendance at pain management seminars. They are accused of exploiting its popularity with those and other tactics and Purdue

Pharma was in the news weekly, if not daily, before the coronavirus pandemic hit the front page.

Schedule 3 drugs also have medical use and less tendency for abuse in the marketplace. Tylenol with codeine is on that list.

Schedule 4 and 5 drugs have little tendency to be abused in the marketplace.

Sally sat quietly as she finished her lunch. She looked at me, but then glanced at the other students seated around us before beginning her own story, "We started partying in college. I was doing oxy-cotton [OxyContin] and pretty soon, the drug got its hooks into me." She kept looking at me directly and it was uncanny because although other students had similar stories, no one interrupted her. I never learned what Sally trained for in college or what she did for a living. The other students listened intently and nodded affirmatively as though their own stories were like hers. I regret not staying in touch with my students because my life got in the way before I retired and began writing.

"The drug got its hooks into me," is synonymous with, "I became addicted to the drug."

"How can I keep my brother from doing what I did," someone else asked, finally. It was a counselor's question, trying to keep a loved one from heading down that dead-end, Addiction Street. Keeping a loved one from heading down Addiction Street is a popular subject for all parents, siblings, spouses, and partners. Sally looked down at her food and picked up her fork again, grateful for or relieved by the interruption.

That was more than twenty years ago. Today drug abuse and death from drugs are even more ominous because opioid use has increased to the point that as many as seventy thousand Americans died in 2020 from it, the death toll every year is in multiples of ten thousand and rising annually. In 2021, overdoses death reached 100,000, yet in 2022, death from overdose was 108,000 in the U.S. The entire coronavirus pandemic had a finite beginning and the end may be in sight. Not so for opioid dependence. In fact, I'll discuss how opioid dependence continued to increase during the coronavirus epidemic.

So far, however, I've talked around addiction without facing it head on. How is it different from my consuming an entire bag of potato chips simply because I opened the bag? When does a bad habit become an addiction, or does it?

The big feature of addiction is that, by definition, patients are physically dependent on a drug. They can't function without it. It's also known as chemical or drug dependence, or more formally, substance use disorder (SUD). It differs from eating potato chips because of that physical dependence. Eating potato chips is a habit, although I wouldn't be insulted if you called it my obsession. I like pretzels too but if you took away my potato chips or pretzels, I wouldn't have a physical withdrawal reaction. I might drink something, chew some gum, and weigh myself because eating them, as I said, simply results in thirst and weight gain. Opioid use results in drug addiction, and if the opioid is discontinued, the result is withdrawal, what recovering addicts refer to as feeling lousy, or sick. As soon as opioid users begin looking at the time, anxious about their next dose, addiction has begun to set in.

"Nearly 44,000 Americans a year – 120 a day – now die of drug overdoses," the New York Times reported in a 2015 article entitled, *Heroin, Survivor of War on Drugs, Returns With New Face*. The article also went on to give that death rate some perspective. In the USA, that year saw more deaths from opioid overdose than those caused by guns or even traffic accidents. Think about that because as I began writing this early chapter, a presidential election was gaining momentum and the incumbent candidate was running on an anti-immigration platform and the coronavirus pandemic was just beginning. The subject has begun picking up steam that the USA ought to ban assault weapons although statistics tell us with more people dying from opioid overdose than from guns, we would be better off running for office on an anti-opioid platform. Maybe I should run on that platform. The assault weapons debate rages still, aggravated by mass shootings, as death from opioid overdose continues to rise.

Both the human and fiscal costs of addiction are very high. To see how deeply into society this issue has burrowed; all we should do is look at a few more statistics. What follows is a quote from a *SAMHSA (Substance Abuse and Mental Health Administration)* report.

"In 2011, *DAWN (Drug Abuse Warning Network)* estimates that about 2.5 million ED (Emergency Department) visits resulted from medical emergencies involving drug misuse or abuse, the equivalent of 790 ED visits per 100,000 population. For those aged 20 or younger, the rate is 500 visits; for those aged 21 or older, the rate is 903 visits."

That is a lot of emergencies from opioids alone, 2.5 million of them in one year. Even more than a decade ago, the incidence of admission for opioids was as much as seven times higher than it was for broken ankles.

The public health threat from opiates and opioids is complicated because abuse of prescription opioid drugs is every bit as common as abuse of illegal drugs. The U.S. Centers for Disease Control estimates on their website that "Every day 44 people die as a result of prescription opioid overdose."

On December 18, 2015, Gina Kolata reported, also in the *NY Times* that, "twice as many Americans died from drug overdoses in 2014 as in 2000." Drug overdose deaths doubled in fourteen years. Ugh!

And it continues to grow. More recently, opioid overdose deaths in the U.S. have been reported to reach more than 100,000 per year.

It reinforces what I have already written that drug abuse leading to addiction is widespread, dangerous and expensive and deserves more of our attention. Some of us shrug and say, "I'm not surprised," and the rest of us shake our heads and say, "How horrible." Some people still look grim and say, "They need to go to jail," a position I am against because drug addiction is a medical issue, not primarily a legal one.

We all agree, however, there is a big drug abuse problem that consumes a lot of resources, both financial and medical, not to mention causing a massive loss of lives. It is a big public health problem and a tragedy.

Although I view opioid dependence as a medical issue, drug dependence differentiates itself from other illnesses in other ways too. One of those differences is a feature called tolerance. I describe tolerance as a drive that forces people who take opioids and other dependence-producing drugs, to continually increase their doses each time they take the drug because the drug seems to lose its effectiveness with repeated use. That's called tolerance and I can't put a time limit on it, but some users complain that after a week of one or two tablets daily, they suddenly find themselves looking at the clock as they wait for their next dose. Tolerance is so dramatic that it is considered one of the differentiating characteristics of drug dependence, if the patient must increase his or her dose to achieve the same effect as the last dose, that patient is developing dependence on that drug.

Finally, there are other features of opioid abuse and the first is one that all users know yet avoid, namely that taking opioids and other drugs will cause them harm in some way. There's danger that users will overdose, that their drug of choice will be contaminated with some other drug like fentanyl or its derivatives. Its presence in both heroin and other drugs has become common, or that they'll use up all their resources buying their drug of abuse, and to buy more, they'll agree to sell the drug, until they are arrested as a drug dealer. The motivation for users to continue taking opioids is stronger than their judgement to avoid them and that is one of their well-known and inevitable negative consequences.

The 1960s was a time of experimentation with drugs, and in a "which came first" paradox, we look back and say, "Drugs caused huge social change," while others reverse the cause and effect and say, "Drugs accompanied huge social change." I win the Boomer award because I witnessed it first-hand having gone through both high school and college in that decade and I had the same social experiences as the rest of my generation. I saw the movie *Help*, starring the Beatles and listened to Sonny and Cher's version of Dylan's "Like a Rolling Stone." My white beard and salt and pepper hair belie my age.

The middle of that decade also dramatically changed how we viewed and treated opioid addiction. That was when addicts were introduced to methadone and the well-known expression *methadone maintenance* entered our vocabulary.

The August 23, 1965, an issue of *JAMA*, (*Journal of the American Medical Association*), ran an article by two researchers, biochemist Vincent Dole and psychiatrist Marie Nyswander, working at what is now called Rockefeller University in New York City. They introduced America to their paradoxical treatment for heroin addiction they named methadone maintenance. It would become controversial. It's paradoxical because methadone is an opioid drug, used to treat opioid dependence.

Their article was titled, "Medical Treatment of Heroin Addiction" and in the abstract of that article they wrote, "With this medication, and a comprehensive program of rehabilitation, patients have shown marked improvement; they have returned to school, obtained jobs, and have become reconciled with their families."

It's difficult from our perspective more than half a century later to see what a reversal of our value systems their medical report represented. According to extant U.S. law, until their *JAMA* article, treatment of drug abusers was based on the 1914 Harrison Narcotics Tax Act that specifically made opioid dependence a crime along with the use of any opioid for the sole purpose of maintenance, which meant, before methadone, any drug given solely to avoid withdrawal. Dole and Nyswander changed everything. Please note that in the abstract of their article I quoted above, they began calling their study subjects patients, not addicts. It was the very beginning of treating opioid addiction as a medical condition and not treating opioid users as criminals.

In the half century that has elapsed since their report, methadone has emerged as one of the standard treatments for heroin addicts in the U.S. But we must look at it a little more deeply to understand it.

As part of Dole's research, he resurrected methadone from history. It was a German invention, synthesized during World War II and introduced by the Nazi regime decades before as a painkiller. It was specifically an opioid designed and synthetized to take the place of morphine because Germany, as part of their wartime strategy, tried or was forced to make itself independent of international trade. Methadone was a German product, manufactured by E. Merck, originally a German company although, I'll discuss later, the original parent of the U.S. company Merck.

Perhaps Dole said, twenty years after World War II, "Let's look at methadone as a treatment for heroin addiction."

I'm certain at least one of his graduate students responded, "What's methadone?"

It began as a product of the I.G. Farben concern, itself an amalgam of dye manufacturers put together way back at the end of World War One. Even the name 'Farben' is a play on the German word 'Farbstoff,' which means dye. Some of the better-known German companies amalgamated to form Farben, including my former employer BASF, along with Bayer, AGFA and Sanofi. During World War II, I. G. Farben became one of the largest contractors to the Nazi regime and is best known, so to speak, for its introduction of Zyklon B, the agent of mass assassination in the concentration camps. In one of history's frequent ironies, a Jewish chemist named Fritz Haber, originally synthesized Zyklon B as a pesticide.

After World War II, the assets and intellectual property of I.G. Farben were appropriated by the allied powers and methadone was taken over by the U.S. pharmaceutical company Eli Lilly. They branded it Dolophine. I.G. Farben was disbanded and broken up by the German government. I worked for a subsidiary of BASF.

The paradox Dole created was his proposal to substitute methadone for morphine as a treatment for that addiction. What attracted him to look at it and to head into the wind like that? His novel idea was to treat addiction as a medical issue, a disease, not as a crime, morality statement or a legal transgression. It was the first time that anyone proposed that interpretation. Dr. Dole teamed up with Dr. Marie Nyswander who worked in his lab and their partnership became famous. It was successful scientifically for their use of methadone to treat opioid addiction, and romantically because they were also married. Sadly, Dr. Nyswander died in 1986 and Dole outlived her by twenty years.

Methadone maintenance does not halt drug use by the addicts; it simply substitutes a measured dose of a prescription opioid for a street drug. The methadone "maintains" the patients during their illnesses by preventing drug-seeking behavior and blocking withdrawal. Methadone has that added benefit of eliminating drug-seeking behavior because the methadone-maintained addicts know they will not get "sick" in spite of not taking their "regular" street drugs. "Getting sick" is another bit of street talk, like feeling lousy, for entering withdrawal.

On the surface, that doesn't seem to make sense, but we are forced to look at the reasoning and the results. Dole's reasoning started with his firm belief that addiction was a medical problem, not a crime, and should be treated medically. As a result, his idea not only removed addicts from the illicit heroin trade but brought them into the clinics. By itself, it did not create ex-addicts, but it fulfilled Dole's vision of changing their status from criminals to patients. I already said it maintained them that way. Until the program was established, heroin addicts were treated by the public and news media as people of lower character, morally unfit and headed for prison because drug addiction was a crime under the law. Methadone changed opioid addicts' status from criminals to medical patients under treatment for a disease. It produced sociological changes as a result.

Addiction also supported racial profiling because back in the 60s and 70s, addiction was mainly seen in financially disadvantaged people and Black ghettoes. Sociologically, it was before drug addiction began to spread to the more advantaged classes.

It worked simply. Methadone is twice as potent as heroin; so, one-half the dose compared to heroin is sufficient to satisfy our body's induced need to avoid opioid withdrawal. Its onset is slow, so there is no "high" or rush, unlike heroin, which has a quick onset that produces a "rush," also known as a buzz, and about three hours of effect. Methadone's effect lasts a full 24 hours. That blocks any high from heroin at its usual dose, so methadone not only blocks withdrawal, but it also prevents addicts from obsessing over their next dose.

Similarly, using heroin itself in a similar medical setting wouldn't work as well as methadone. The pharmacological advantages of methadone's higher potency, longer duration, and lack of buzz, or high, still give methadone the advantage.

The program was a brilliant idea, but it turned out to have a controversial downside. Methadone on its own has become a street drug and it's a lot more toxic than heroin for the same reasons that makes it useful to treat addiction, because it lasts a full day and is twice as potent as heroin. As a result, many methadone abusers end up in the hospital from the methadone and not from heroin. It even has familiar street names like Dollies, Jungle Juice, Wafer, or my favorite, Chocolate Chip Cookies, a name derived from clinics administering methadone by sprinkling it onto cookies.

I can't report on my addiction-counselor-in-training students' successes because travel for work dominated my life for a decade. I had to give up my Pennsylvania job as my company began negotiations to sell the division that employed me. My headhunter found me the same job managing drug development but for a German company that required my presence in Germany and the U.K. every other week and California monthly. That schedule didn't leave any time to do my laundry or even have a date, let alone to teach. The German company even bought my

Pennsylvania farmhouse from me so I wouldn't have to spend time selling it. I could concentrate all my energy on managing their international drug-development teams.

Obviously, I've followed drug addiction in the U.S. and Europe closely and a recent HHS report taught me something fascinating. Increased opiate abuse in the U.S. was limited to poorer and more-rural areas for decades but has now begun to spread to middle- and upper-class areas as well. Whereas the drugs became a substitute for the stress of economic opportunities among less advantaged populations, as the sociology of drug addiction changed, it spread to more advantaged populations for reasons other than as a substitute for economic opportunity. People use it to counter business, travel, or family stress. It used to be considered an urban ghetto problem, but its use has become so widespread that it is has become a general, population-wide problem. I'll show later, it has also spread to Europe.

Drug addiction is recognized and accepted as a medical problem but now we can add that it has also become a sociological problem. It's still challenging us fifty years after we began methadone maintenance and opioid use is still spreading.

The Jersey City State class I was teaching was beginning to squirm and I glanced at my watch. It was almost three PM and although I still had their attention, my feet were beginning to hurt and I wasn't looking forward to driving back to Pennsylvania, so I asked, "Any additional questions before we close?"

One of the class members said, "Will you teach any other course during our certification?"

"Unfortunately, not, but I've enjoyed today. If you have any questions, my email address is on the board. Thank you all."

That class gave me dramatic memories of former addicts hard at work helping others, like what I, a former teacher, attempt to do writing about addiction. I said, "Thank you all."

They applauded and it was over.

I must miss student contact because I still enjoy talking about it.

I've read that mental illness and opioid addiction too frequently co-exist in the same patient. The thinking is not that the two conditions correlate, as I discuss elsewhere, but that mental illness and opioid abuse are due to the same youthful trauma.

Mental illness is our reaction to that trauma, rendering us unable to function as adults because we are simultaneously wrestling with the pain, including but not limited to loss of a parent, abuse by a parent, dysfunctional family life, parental or familial drug use, while we wrestle with the tasks of daily living. Drug abusers try to blunt the psychological pain they carry by eliminating all feeling with drugs. It strikes me that perhaps the two diseases even lie along a continuum.

I read about a Colorado psychiatrist, Dr. Paula D. Riggs who established an organization she's called Encompass: Integrated Mental Health/Substance Treatment. She treats mental illness and drug addiction simultaneously, seizing on the idea that the two diseases co-occur.

It's not as morbid as it seems.

About 60% of addicts, SUD (Substance Use Disorder), patients exhibit mental illness at the same time. Depression, Generalized Anxiety, Bipolar Disorder and PTSD (Post-Traumatic Stress Disorder) are a few of the common ones. The flip side of that is about 60% of mental health sufferers also abuse drugs. Those figures are especially true for adolescents, a scary thought for those of us with teenagers. It also hints at another sociological concern, that opioid abuse is more common among young people.

So, we're left metaphorically scratching our heads asking, "Does one cause the other?"

I've stressed a few times, correlation does not mean causation, so we're forced to conclude mental illness does not necessarily cause drug abuse nor does drug abuse cause mental illness. In a later chapter, I'll discuss how the murder rate and ice cream consumption both go up in summer, but one doesn't cause the other. They both increase due to the warmer summer weather.

Both mental illness and SUD are caused by some third factor, a few of which I've already mentioned, childhood trauma, parental abuse, incest or rape, unstable family, familial drug use, injury, and generalized anxiety disorder. Having a youthful problem like one of those and spending the rest of one's life trying to outrun it, especially true for sufferers of ADHD (Attention Deficit Hyper-Activity Disorder).

What are we to do?

Get help or offer help, earlier rather than later. Don't be afraid to intervene.

Treat childhood ADHD, or, if you recognize it in someone else's child, bring it to that child's parent's attention. If you witness child abuse, call the cops or child services, it's a public service. If your family life is unstable, don't tolerate it, get help. If it's beyond help, separate. It's not your partner's fault you can't get along, it's nature… human nature.

To quote the National Institute on Drug Abuse website (NIDA.NIH.GOV), "Although drug use and addiction can happen at any time during a person's life, drug use typically starts in adolescence, a period when the first signs of mental illness commonly appear."

I continually stress, don't wait for someone else, lead the charge. Speak to parents, call authorities, help the student seek help or talk to your partner. As Peter, a recovering addict I introduce in a later chapter told me, leave your ego outside, along with your judgmental opinions. Then offer to help.

The presence of two diseases, such as mental illness and drug addiction, is known in medicine as a comorbidity.

Notwithstanding the name, it's not as morbid as it sounds.

There's a frightening side of SUD and Mental Illness comorbidity. A medical journal (*BMC Psychiatry*), published in 2019 corroborates what I said above. The authors, researchers in Sweden and Norway wrote, "Most patients had a family history of social disruptions in childhood, destructive home conditions, psychiatric disorders and persistent abuse of alcohol and drugs." Then they concluded from their study of more than a thousand patients, more than half died prematurely compared to the general population.

So, the presence of a mental illness and opioid abuse and addiction, while not necessarily a death sentence, raises the risk of early death.

As I said, a scary concept.

Their article also demonstrates that when opioid addiction spread out of American Urban ghettoes, it didn't stop when it spread to more advantaged populations, it spread all the way to Europe. That *BMC Psychiatry* article said it best, "A study in nine European countries reported an overall mortality ratio of SUD patients abusing illicit drugs that was 10-20 times the level in the general population of the same age and gender."

It's not just a scary concept, it's a lethal habit.

Closer to home, deaths from drug overdose have also begun to mirror older racial divides. It's almost paradoxical that the drug addiction epidemic spread from less advantaged to more advantaged populations, but it also began to teach us that Black adults are more at risk than White adults.

I base that statement on a report, published in 2020, from the U.S. Centers for Disease Control and Prevention, entitled, *Drug Overdose Deaths rise, Disparities Widen.*

Disparities? I quote the report, "In counties with more income inequality, overdose deaths for Black people were more than two times as high as in counties with less income inequality in 2020."

That same year, "Overdose death rates in older Black men were nearly seven times as high as those in older White men…"

What's going on? Why do opiates kill more Black men than White men?

One of the theories the CDC puts forward is availability of health care. Perhaps it's better among the White population than the Black population?

The CDC report subtilty argues against that with the statistic, "Overdose death rates for younger American

Indian and Alaska Native women were nearly two times those of younger White Women." Assuming the Indian Health Service offers the same level of care as private health care, why do drugs kill more Native American women than White American women?

Perhaps it's the availability or frequency of emergency medical care. That could be studied because that could be measured.

Are more drugs contaminated with illicit fentanyl associated with Black deaths? That could also be measured.

U.S. public health is challenged by this apparent disparity in deaths from drug overdose that resembles higher education in the forties and fifties in the U.S. In a modern society, that seems finished with the stupidity of exclusion by race, dogged by a public health issue with no single cause.

It's more complicated than simple health care delivery or availability. The CDC reports separately that diabetes prevalence among Black people is 11.5% and 7.7% for Whites. Diabetes is more genetically driven than illicit drug use suggesting, if we can draw any parallel conclusion from those figures, that differences in death from drug overdose may be more a physiological function and not a sociological function.

Something is going on and we don't understand it very well. It needs more study.

Plant Juice

Ancient biblical history teaches a fundamental tenet, that the Jewish patriarch Abraham was born in Ur, the ancient Sumerian city that is now part of Iraq. Historians tell us those Sumerians formed the first modern civilization in the region we also refer to as Asia Minor. It is also called Mesopotamia. We learned in middle school that Mesopotamia translates as "The Land Between the Rivers," and those rivers are the Tigris and Euphrates.

One theory of Sumerian origin is that they were among the first modern humans who migrated out of Africa so when we say Sumerians have lived in that region a long time, it has been a very long time. Archeological evidence teaches us that Sumerians settled the area at least 5000 years ago and the region in which those rivers meet, their confluence, is also known historically as Babylon, where the biblical story of Noah took place. That's where legend says he saved all the animals from a flood and the flood was apparently the high-water associated with that confluence of the Tigris and Euphrates rivers.

Some of the physical evidence archeologists uncovered include 5000-year-old Sumerian clay tablets that describe raising poppies and preparing opium. Five thousand years is well before the Old Testament was put together, back to the very beginning of written history. It's not only our earliest record of poppies, it's one of our earliest purposefully written records of anything because Sumerians also gave us handwriting, in the form of cuneiform, to preserve thoughts. We've evolved from moist clay tablets to pencil and paper, to typewriters, to laptops and cellphones with voice recognition as we struggle to preserve thoughts without trying to read our own handwriting. Mine has always been illegible and my thirteen-year-old's is too.

Throughout opium's long history, the rush it produces is one effect the crude drug consistently engenders. I also call it the buzz, or the high. Although that rush still surprises many people, it is part of the reason opium has been with us since before the Bible was written. To this day, people enjoy and seek that rush. In some circles, as I mentioned, users call it a "buzz," and, whatever they name it, they invariably describe it as pleasant and welcome.

Poppy plants are also the same flowers we have also associated with death and burial since as far back as the Egyptians. John McCrea wrote a poem in 1915 that begins, "In Flanders Field the poppies blow..." His poem commemorates a World War I battle in Belgium and points to why poppies are the flowers handed out on Memorial Day to remind us of fallen heroes. It is all so peaceful and civilized.

Normally, after the plants produce their pretty poppy flowers and attract bees to do what they do, the flower petals fall off and the plants appear to stand at attention as all the green seedpods point straight up. Those seedpods appear as images in classic early sculpture because poppy plants are programmed by their genetic makeup to wait, so to speak, for their seedpods to ripen, dry out, turn brown and burst open to let the wind spread poppy seeds that will sprout in the spring to produce the next crop. Humans have intervened over the years to harvest those ripe seeds and use them as a baking spice, in poppy seed bagels for example, or my late maternal grandmother's delicious poppy seed cookies.

But occasionally, some people have ideas that have nothing to do with the taste of poppy seeds. Those in the drug trade have learned to slice the seedpods slightly while they are still green, called lancing the pods, and then wait, usually overnight and sometimes longer. I show below, the pods exude a white, latex-like material from the lancing and that white goo dries in the sun. The material the plants leak out is known as poppy tears in some cultures and my photo can be interpreted as a picture of poppy pods crying.

We call those poppy tears raw opium.

We've given opium poppies the Latin name *Papaver somniferum,* and those Sumerians discovered how to harvest those poppy tears. You can see how the plants ooze a milky substance, as we see in the photo above, that milky substance is known as raw opium. We don't have a name of who discovered that little trick of scoring the seedpod and collecting the milky substance, but poppy growers still do it that way.

Poppy cultivation spread from Mesopotamia south to Egypt and east, along what became known as the Silk Road, to China and India. It's unclear if Marco Polo ran into opium trade during his Chinese sojourn but either way, opium was traded all along the Silk Road. China has announced a big project to redefine the old Silk Road for modern commerce. It's unclear what that will do to the ancient opium trade or even if they will identify it, let alone, own up to it.

Opium and products based on or designed around chemicals isolated from that ancient Middle Eastern plant have developed an international market among people who otherwise would never have met or done business with each other. Innovating around the chemicals in the plant has spread throughout the world. Those chemicals include morphine, codeine and papaverine, and they have given us our modern chemical and pharmaceutical industries that employ tens of thousands of people and markets synthetic opioids to hundreds of thousands of them. But those same chemicals have also required us to fund well-armed police forces and support a population base of sick people who all seem to repeat the same mantra, "I want to stop using this stuff, but I can't."

I always say, the drug grabs a hold of you and won't let go. That's addiction.

Flashing back just 2000 years, Egyptians even used opium as a teething remedy for children, among other uses. I'm certain it worked well although their society didn't survive nearly as well. Maybe too many Egyptian children teethed on opium seedpods.

Even more recently, the Greeks wrote about opium. Homer mentioned it in both the *Iliad* and the *Odyssey*. He wrote, in a translation of his Greek original *Iliad*, "… help me to my black ship, and cut out the arrow-head, and wash the dark blood from my thigh with warm water, and sprinkle soothing herbs with power to heal on my wound whose use men say you learned from Achilles…" His "soothing herbs" undoubtedly were opium poppies.

Ceres, the Italian grain goddess after whom cereal is named, took opium to soothe her pain. I don't know what hurt her but I'm sure the drug cured whatever it was, or at least she didn't complain about it anymore.

Arab physicians used opium and Arab traders introduced it to China around 800 AD. That makes the stereotypic Chinese opium den an Arab export although we'll see that Britain also played a role in its spread through China.

The word opium comes from the Greek word *opion*, which translates roughly as plant juice. My photo shows how that very juice squeezes out of the fruit of the plant known as opium and tempts us to repeat the adage, "The Greeks had a word for it." The drug morphine was named for Morpheus, the Greek god of dreams. Given that the

Greeks named the plant and its main product, it makes sense that opium production spread quickly from its origins in Asia Minor, to ancient Greece. There are third century BC Greek references to poppy juice. That writer called it *meconium*, a word we still use today although now we use it to describe the first stool passed by a newborn baby. It is unclear how opium found its way to ancient Sumer and Egypt, but it is safe to conclude that the opium trade has also been around for at least as long as people sought that poppy juice.

It's easy to get lost in who used opium and how long ago they did so, without really understanding what it is or how it works.

Sumerians were a very creative and innovative people although it is hard to appreciate that today because of regionally unstable governments, fractious religious relationships and seemingly endless war in the region that today includes modern-day Iraq, Iran, Turkey, and Syria. Turkey is the model of stability in a region not known to be stable.

The Sumerians developed features of modern civilization we take for granted, including, but not limited to the wheel that led to the invention of the donkey cart and that evolved directly into bicycles, trains, streetcars, automobiles, and trucks. They also began to plant and harvest cereal crops, such as wheat, and their success at crop cultivation also gave it the nickname, the Fertile Crescent. As I mentioned, they invented handwriting in the form of a text we have labeled cuneiform because they used a pointy object to mark little triangles in soft clay and those triangles are wedge-shaped. We can still view their messages today in museums, including how they used symbols for counting, the earliest mathematics. Historians tell us Sumerians also invented the arch and were the earliest astronomers because they began looking at the night sky. That led directly to Galileo and the science of astronomy. Not bad for folks who also began, about 3,400 years ago, to cultivate opium poppies, along with their wheat. Then they traded poppy seeds to the north, all the way to Greece.

Moving ahead in history a thousand years or so, Britain colonized India, first through their British East India Company followed by what is called the British Raj from 1858 to India's independence in 1947. *Raj* is Hindi for "rule" and the root word for *rajah*. During their domination of India, British merchants traded Indian opium for Chinese tea. That not only led to rampant opium addiction among Chinese, but it also led to Britain fighting two opium wars with China. The widespread use of opium in China turned out also to be around the same time Chinese workers began coming to America. Those new Chinese immigrants carried their opium with them.

According to the U.S. DEA Museum, those nineteenth-century Chinese came to America to work in the California Gold Rush and on the railroads as train service expanded rapidly across the country until finally establishing a rail link tying the west coast to the east coast in 1869. That sped up the spread of opium from California east to New York, perhaps an unexpected side effect of establishing a rail linkage. It's amazing how opium followed the spread of sociological and cultural change.

But until the 19th century, all users had only raw opium. Left in open air, the milky white substance the poppy exuded dried and turned brown and people either smoked it or chewed it. There was chemistry involved in figuring out that raw opium was really a mixture of drugs.

I'll explain how in later chapters.

So, a plant from what we now call the Middle East, spread throughout the world because people liked what the plant produced, but they were limited to smoking the weed or somehow taking the white resin.

Either way, they sought the same buzz modern users seek.

Morphine, Dyes & Aspirin

The Birth of the Pharmaceutical Industry

W ay back in 1805, a German pharmacist, Friedrich Wilhelm Sertürner succeeded in extracting morphine from opium. He showed us that morphine had the same sleep- and dream-inducing properties as did opium and concluded, correctly, that opium's effects were due mainly to its morphine content.

It would be incomplete if I failed to mention Sertürner's transgression in testing his pure morphine, he tested it on a group of seventeen-year-old boys and on himself. We look at that with modern morality and ask, "How could he do that?"

There's no real answer but seventeen-year-old boys or girls don't belong in an opiate clinical trial.

Nevertheless, Sertürner gets credit for being the first person to isolate a pure drug from a plant source and his discovery is one of the factors in the growth of a modern, recognizable pharmaceutical company.

That pharmaceutical company began when a German family named Merck bought the Angel Pharmacy in Darmstadt, Germany in 1688 and they eventually called themselves E. Merck and Company. After nearly a hundred fifty years of compounding drugs, by 1830, E. Merck began manufacturing Sertürner's morphine and transformed itself from an old family pharmacy to subsequently, a pharmaceutical manufacturer of synthetic drugs.

But the company also became embroiled in politics and war through global expansion. Let's see how that happened.

In 1891, Georg Merck, one of the brothers who ran the company, came to the U.S., and established a subsidiary of E. Merck in New York. He named it Merck and Company but a year before the Armistice ended World War I, in 1917, the U.S. government confiscated the American branch of Merck and Company under the Trading with the Enemy Act of 1917, along with a host of other German companies, so today, there are two companies named Merck. One is the U.S.-based big-pharma company that calls itself Merck and the other is the German company that calls itself E. Merck, Darmstadt. They are not related, despite their founding by the same family.

Morphine transformed E. Merck from a family that owned a pharmacy to two global companies that synthesize drugs.

A generation after Sertürner isolated morphine, Sir William Henry Perkin was another Englishman who took his first job at age fifteen. It was not uncommon to begin work that young in the 1850s, yet what he accomplished still makes me shake my head and ask, "How did he pull that off? Precocious young man."

He went on to change the way we view the world, and he did that while searching for something else.

Perkin was apprenticed to a German chemist, August Wilhelm von Hoffmann, who had moved from Germany to England by invitation to start the Royal College of Chemistry. Von Hoffmann was recruited for that post because he was eminent, having studied under Justus von Liebig.

High school chemistry students know von Liebig because his name is attached to his "Liebig Condenser," used around the world to condense gases into liquids. I was introduced to it in high school chemistry class, and it is a simple contraption with a cork-screw-shaped glass tube inside a larger glass tube. We circulate cold water through the larger tube to cool the corkscrew and that causes the gas inside the corkscrew to condense and drip out the end. When a Liebig condenser is made very large, we call it a still, the kind people use to make moonshine. It has been a very useful contraption ever since Liebig thought of it.

Perkin was a precocious East-Londoner in 1853 with a penchant for chemistry when he joined von Hoffmann as a research assistant, but he also had a bright teenager's curiosity. Von Hoffmann assigned him the project of developing synthetic quinine and he wanted Perkin to use coal tar as a starting material.

That period at the end of the Industrial Revolution was well before our modern era when we began to "turn the lights on," with a wall switch, let alone begin brewing coffee and seeing who's at the front door with cellphone applications. Homes were lit by gaslight at night and companies made the gas fuel for those lamps from coal. They heated the coal until it released what they appropriately named coal gas and they collected that gas and sold it so their customers to burn for light. Heating the coal to harvest the coal gas left companies with a black sludge they called coal tar and it was possibly the first industrial waste.

Von Hoffmann's idea was sound because quinine had been a natural product up until then, extracted from the bark of the Peruvian *Cinchona* tree or *quina*, which gave it its name, and it was an important product because it prevented malaria.

Many British colonies were in tropical climates in the mid-nineteenth century, the peak period of Britain's colonialism so they were big consumers of quinine because tropical climates are rampant with malaria. It would have been a coup if Perkin could have done it with coal tar because that would have made quinine the first drug made from anything other than a seed, flower, plant or tree, and coal tar, as a byproduct, was cheap, plentiful, and ripe for utility.

Extracting quinine from *Cinchona* was also expensive so whoever invented a synthetic substitute could stand to make a lot of money too, creating a double incentive, but Perkin's experiments didn't go well and when von Hoffman was out of the lab, Perkin let his own ideas creep into the carefully regulated scientific environment. It was the Victorian era, after all and he was clearly incorrigible.

Perkin chose aniline for his little side project. It was a chemical discovered twenty-five years before by another young German chemist.

Aniline is also the single, unexplored chemical that led to the birth of entire dye industry. The irony of it is that aniline also started as a plant product, from another young German chemist who distilled indigo plants and that yielded a substance he named *crystallin*. Other researchers after him isolated similar substances and several of those substances turned blue when the researchers treated the isolates with lye or other caustic chemicals. Someone named the blue products aniline, after a plant called *anil* that, as another source, could also be distilled for its indigo. Perkin's mentor, von Hoffmann taught us that all those blue isolates were the same molecule, aniline.

Perkin substituted aniline for the coal tar he was using in his experiments, and he tried to extract the aniline with alcohol. It turned the alcohol purple and when he dipped a piece of silk into the purple alcohol, the silk took up the purple dye. He called the dye *mauve* and he had inadvertently invented the first aniline dye. Neither he nor anyone else ever did manage to invent cost-effective synthetic quinine.

He was only 18 when he filed a patent application for the dye and his father saw his potential and trusted him sufficiently that they set up a business named Perkin and Sons. Within two years of the patent, they began commercially manufacturing the *mauve* dye and they sold it under the trade name Mauveine. Within five years he was a wealthy man and within a decade, the dye industry had moved across the Channel, as the growing industry centered itself in Germany, Switzerland, and Austria. Perkin gets the credit for giving us the entire class of chemicals we still call aniline dyes, and we'll see that aniline dyes led us to pharmaceuticals because the underlying process of creating dyes from common chemicals is the same process that chemists use to create drugs from the common chemicals.

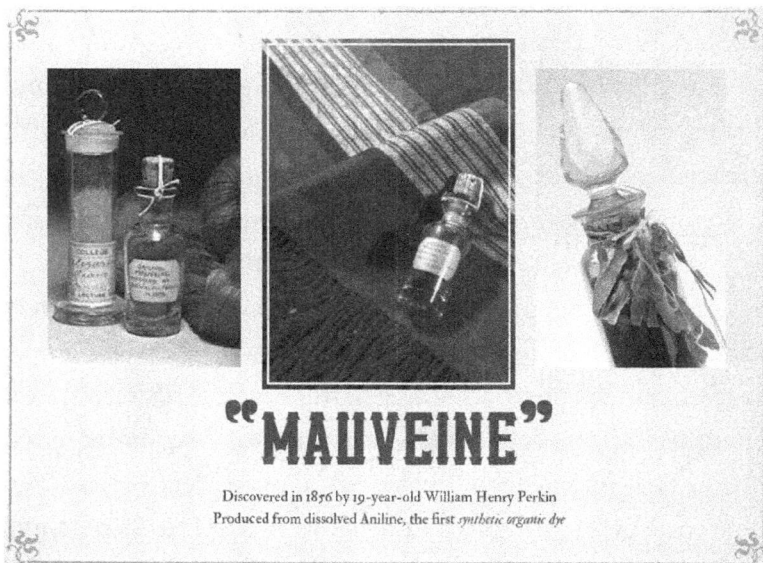

"MAUVEINE"

Discovered in 1856 by 19-year-old William Henry Perkin
Produced from dissolved Aniline, the first *synthetic organic dye*

A mechanical invention also played a role in opiate history in the mid-nineteenth century. The routes of drug administration expanded profoundly when a Scottish physician, Dr. Alexander Wood invented the glass syringe and hollow needle and began using his invention to inject morphine into his patients. He always claimed he was inspired to invent the needle and syringe by thinking about bee stings because bees use their hollow stingers to inject bee venom. Whatever his inspiration, he gave us the word *subcutaneous,* and he described his invention as allowing subcutaneous administration. Inspiration from nature.

His invention has a sad side story though. Wood injected his wife and she died from morphine-induced respiratory depression. It was possibly the first death from an injected overdose…and it was administered by the guy who invented drug injection.

The timing was right though for pure morphine to find widespread use during the American Civil War. In fact, during and after the War, estimates as high as 400,000 soldiers were addicted to opium. It was so common that opium addiction was referred to as the "Soldier's Disease" and during reconstruction, it wasn't unusual to see veterans with leather bags of morphine hanging around their necks and glass syringes with needles. Those handy leather bags were given to them supposedly for pain, as part of their discharge from the military, although we can look back and think if they were addicted to opium during their military service, they were simply enabled to feed their addictions after service. It's also a case of criticizing past actions with modern morality, also an error.

But I implicated dyes and aspirin leading to the pharmaceutical industry and we've looked at dyes. Let's look at aspirin with a side glance at my own career path.

BASF is a German company that began life as a dye manufacturer, known widely a technical generation ago for their brand of videotape, and now as the largest chemical company in the world. BASF is an acronym for Badische Anilin und Soda Fabrik, which translates loosely as Baden's Aniline and Soda Factory. Many of us have heard of the town Baden Baden which helps define the geographic center of the region.

BASF is also the parent company of my former employer. But, back to aspirin.

Before I went to work for a BASF subsidiary, I worked for a big American pharmaceutical company and I was assigned to lead a delegation from our Main Line Philadelphia headquarters to Dresden, Germany to visit a company called AWD (Arzneimittelwerk Dresden or Dresden Medicine Factory) to co-chair a discussion about licensing one of their new drugs from them. As I walked through a lobby area to that meeting, I saw a brass plaque on the wall and I stopped to read it because I'm a brass-plaque junkie, forgive my description, and I can't walk past one without reading it. My German was adequate. The plaque commemorated their discovery of what we now call aspirin.

I'm glad I stopped to read it, but my German host was waiting for me to begin the meeting and he was clearly impatient, despite my being on time. "*Guten Morgan, Herr Gold*, let's get started," he said, nodding in my direction because I was the last one to join the meeting. I'm sure under his breath he was muttering, "*Amerikaner!*"

From Dresden, a colleague and I took a train to Prague for the weekend.

A few decades after I left the pharmaceutical industry, I began writing about drugs and found an old business

card from AWD, the company that hosted my meeting so
many years before. I wrote to the fellow whose name was
on the card, reminding him who I was and the company
that employed me way back when. I summarized, "…
can you provide some history about aspirin's discovery as
commemorated by the plaque in your lobby."

He responded, "We are the company AWD Pharma
in Dresden, and one of the roots of our company was the
former chemical company '*Chemische Fabrik von Heyden*'
in Dresden and Radebeul near Dresden. You got to
know a part of this company in 1997 when you visited
Dresden". He sent me a book, in German, documenting
his company's rich history.

I am still amazed that he remembered me, and even
more so, that he was still there when I wrote. I didn't ask
what impression I made but I still have the book and I
still refer to it although my German is rusty. It's illustrated
and that helps.

Friedrich von Heyden founded his chemical company
in Radebeul, near Dresden in eastern Germany in 1874.
Dresden was bombed heavily during World War II and
under East German neglectful rule for fifty years after
the war. Their big church, the *Dresden Frauenkirche*, or
Dresden Woman's Church was still a bombed-out pile of
rocks when I was there although they have since rebuilt it.
Radebeul is midway between Dresden and Meissen, and I
also had images of a side trip to tour the Meissen porcelain
factory. I never made it and I don't have any of their china
in my collection. In fact, all I've collected over the years
has been glassware.

He named his company after himself, *Chemischefabrik
von Heyden* or in English, Heyden's Chemical Factory and he

developed an industrial-scale synthesis of salicylic acid. Von Heyden collaborated with a University of Marburg chemist, who had determined the structure of salicylic acid, isolated from willow trees. He synthesized it, albeit at laboratory scale and then Von Heyden scaled up the synthesis to provide a source for something he could sell. Chemical synthesis of drugs, like synthesizing dyes, is easier, cheaper, and more standardized than extracting them from natural sources so it followed that industrial synthesis of salicylic acid was more desirable than continuing to extract it from willow trees, the garden plant *Spiraea*, or its other natural sources.

Von Heyden made derivatives of salicylic acid and his *Chemische Fabrik von Heyden* began to produce acetylsalicylic acid in 1897. We know acetylsalicylic acid as aspirin, but he didn't, and he named his new product *Acetylin* after the chemical reaction he used to synthesize it.

It might have been a commercial coup, but it turned out to be the horse that came in second because today no one has heard of Acetylin.

Bayer, Germans pronounce it 'Buy-er,' was founded almost 400 miles away in Barmen, Germany. The same year von Heyden started manufacturing Acetylin, Bayer also started making acetylsalicylic acid. Bayer's lab notebooks show the date, August 10, 1897. Bayer gave it the tradename we all know, Aspirin, and they marketed it successfully. They say the name originated from *Spirea*, the common garden bush that is one of the other sources of salicylic acid, plus an "A" for acetyl. The name stuck. Today we would say Bayer and von Heyden had an intellectual property dispute, but then we've all heard of Bayer Aspirin and no one's heard of Acetylin so perhaps it was just Bayer's good marketing.

Bayer's story is that they hired a chemist named Felix Hoffmann, unrelated to Perkin's mentor Von Hoffmann, and assigned him the project of developing techniques to add acetic acid, which is known as having just two-carbon atoms and is the active ingredient in vinegar. Acetic acid and its chemical cousins happily attach themselves to many chemicals, so to see what possible new or improved products might result, Hoffman added acetic acid to salicylic acid, the product of the willow trees named *Salix* or the flowering shrubs named *Spiraea*, as I said, both good sources of it. Salicylic acid from willow bark had a reputation, since the Greeks, as a good treatment for fever and pain. The product Hoffman made, he labeled acetyl-salicylic acid (commonly abbreviated ASA) and it was better for curing headaches than willow bark extractions and more convenient too.

We know acetyl-salicylic acid as aspirin and Bayer launched it in 1899. They created the tradename with an "A" for acetyl and "spirin" for *Spirea*.

August 1897 was a good month for that Bayer chemist because he continued experimenting with the chemical process known as acetylation. That's the name given to attaching acetic acid, as I said, the active ingredient in vinegar, to other chemicals and it was how he made Aspirin, or as the chemical is known, acetylsalicylic acid. Eleven days after he made aspirin, he added acetate to morphine. It was August 21, 1897, and he had other Bayer employees try it and the legend says, they felt heroic, *heroisch* in German, so Bayer gave his new chemical, called di-acetylmorphine, the tradename *Heroin*.

Trying chemical creations also fit the old stereotype of chemists the world over, historically, they all tasted, smeared, or otherwise tried the chemicals they invented. Not so much anymore because of the risk inherent in new chemicals.

It also connects aspirin to heroin because they were invented by the same chemist, with the same process at the same company.

I discuss heroin and morphine in the next chapter.

Morphine & Heroin

Morpheus was a Greek God, and his Roman counterpart was Somnus. They were both in charge of sleep and dreams. Morpheus' lent his name to morphine, the first pure drug isolated from a plant, while Somnus' name is associated with sleep – the first word that comes to mind is its loss, insomnia. I discuss morphine's history in several places, in part because it's so interesting.

Heroin, I remind you, has a German accent, invented by the same guy who invented aspirin, Bayer Aspirin.

Estimates of morphine abuse teach us that 10% of the U.S. population has tried morphine recreationally and that abuse epidemic continues. In the four years between 2004 and 2008, the number of addicts more than doubled in the U.S. and continued upward from there until today.

Morphine and heroin, indeed all the opiates and opioids are well-known to induce beautiful dreams in addition to their tendency to induce sleep. Those dreams we call hallucinations.

Samuel Taylor Coleridge, among other historical figures, is well-known to have abused opium. He wrote in *Kubla Kahn,* with hallucinatory imagery presumably inspired by an opium dream:

A damsel with a dulcimer
In a vision once I saw:
It was an Abyssinian maid
And on her dulcimer she played,
Singing of Mount Abora.
Could I revive within me
Her symphony and song,
To such a deep delight 'twould win me

Until 1817, only raw opium, just dried poppy plant juice, was the opiate available and people smoked it in a pipe or a hookah or chewed it. That dried plant juice was the main drug in the drug trade for its first few thousand years and one of the effects of opium that stood out was its ability to produce sleep. The poppy plant was traded, countries invaded, and sleep created. The Philistines introduced the poppy plant, along with cumin, to the Judean Kingdom 3000 years ago.

As I discussed, Friedrich Wilhelm Adam Sertürner extracted morphine from opium and showed it had the same properties as the opium, namely inducing sleep, so he named it morphium, after Morpheus. He concluded correctly that the sleep caused by smoking opium was due to the morphine in the opium. It was 1805 and he was the first to offer the idea that many plants had active chemicals we could extract, and we continue to do so today. As I explained, morphine's isolation was one of the triggers of the emergence of our pharmaceutical industry.

I take issue with how Sertürner showed his extract's activity, namely he gave it to seventeen-year-old boys and, at the same time, took it himself. Applying our modern morality, what he did was immoral on a few levels.

Nevertheless, he was the first person to isolate a pure drug from a plant source and his discovery formed the basis of the pharmaceutical industry's birth. We even have a modern term for morphine isolated from poppy juice, we call it a natural product, like the caffeine in coffee, the sugar in sugar cane, the atropine in belladonna flowers or the digitalis in foxglove.

I also looked at morphine use during the U.S. Civil War in an earlier chapter. Addiction in the military was underlined again during World War 2 when the syrette was invented by the old E.R. Squibb & Sons. They were my first employer after my college graduation, before I left for Boston to attend graduate school. The syrette was the first introduction of a single-use injection device, loaded with a measured drug dose. After injecting the soldier, the attending physician or nurse pinned the syrette to the soldier's collar as a signal that he had already received a morphine dose. It announced, "He was dosed. Don't dose him again." Sexism was rampant eighty years ago and combat soldiers during World War II were all male.

Big-Pharma is how we describe the pharmaceutical industry (pre-biotechnology) and morphine helped found it.

Morphine induces beautiful dreams.

Either way, the syrette pinned to the soldier's collar announced visually, "Don't dose him again."

Morphine's role in producing physical dependence depends on the way it's used. If it is taken as a recreational drug, users will show tolerance to its effects, begin to dose themselves more frequently and continue using it to ward off dreaded withdrawal symptoms. That dosage pattern reliably and reproducibly causes addiction.

However, if morphine is given to cancer patients to blunt their pain, the patients still exhibit tolerance and physical dependence. The distinction with this population of users is that the patients seldom exhibit recreational or continued use and their motivation is to avoid pain, not seek a buzz.

I found an N.I.H. publication titled *Development and Validation of the Current Opioid Misuse Measure*. There were seven authors listed, all scientists. They write, "Despite international attention to improve pain management, inadequate pain relief is a serious public health issue." They were addressing opiate use in the cancer patients I mentioned as well as non-cancer patients I discuss elsewhere.

They went on to say, "It is also important for the successful treatment of chronic non-cancer pain to be able to frequently monitor patients on opioid regimens and to identify those patients who exhibit ongoing abuse behaviors," but they also conclude, "Opioids will likely continue to play a critical role in the treatment and management of chronic noncancer pain."

So, they focused their research on validating an interview technique to identify chronic pain patients who exhibit abuse behavior. It's a vital, yet complex task that treads the line between kind humanity and amateur enablement.

It also points to the limitations of opiate use. Morphine and other opiates don't treat underlying conditions, only the pain those injuries or diseases cause. It was even sold without prescription until it was classified as a controlled substance in 1914.

Published estimates of morphine abuse teach us that 10% of the U.S. population has tried morphine recreationally.

The 1930's received an exclamation point when, in 1939, Sigmund Freud's physician killed him by overdose at Freud's own request, physician-assisted suicide. The famous psychiatrist was in pain from his jaw cancer, and he elected death as a permanent pain treatment.

I've shown that heroin has been around more than a century and its use was outlawed in the U.S. during most of that time. Despite that, it is still a problem. Today, in the U.S. we call it a Schedule 1 drug, defined by law as having a high potential for abuse and no medical use. How does something that is illegal to import, manufacture or sell and dangerous to take, remain a bestseller, when selling it can result in arrest and using it can kill the user?

One of its street names is, "horse," and it's not a winner, don't bet on it.

First, there is a supply chain that has grown out of control. The bulk of the heroin sold in the U.S. comes from Mexico and South America with a little arriving from Afghanistan although that source is lessening. Clearly, it's being smuggled in, yet as our borders are sealed tighter and tighter, they seem more and more porous. Decreasing immigration from Latin America didn't decrease heroin availability so it was clear those immigrants weren't bringing in the heroin in their luggage. It comes in some other way, usually on the backs of illegal immigrants sneaking across the border in the middle of the night or hidden in the trucks and cars crossing the border.

Heroin is still a problem because there remains a market for it and that market, until very recently, continues to be stimulated by how it works; it's an opioid that produces a pleasant buzz. People who've experienced the buzz, seek it again. I found a paper that said a heroin buzz is love.

The *Drug Enforcement Administration* (DEA) issued a press release in December 2019 entitled: *More than $1.2 million worth of heroin seized; 15 indicted during operation tied to 15 overdoses, three fatal.* That seizure was on New York's Long Island making it clear there's still a large market and the location shows how its use spread to the suburbs.

Heroin fascinates the scientist side of me because of how it works. Users swallow or inject heroin and their bodies immediately convert it to morphine. It shows up in users' urine as morphine. That is also a bit of a head-scratcher, because heroin is twice as potent as morphine, that is, a dose of heroin is half the weight of a dose of morphine, yet our bodies turn the heroin into morphine and users say what is effectively a lower dose, still gives them a buzz. I suspect the buzz comes from heroin's faster onset because it can't come from a lower dose.

Morphine was the first natural product isolated from a plant. Heroin was one of the first fully synthetic drugs made. Morphine is used medically. Heroin is abused.

How Does So Much Heroin and Fentanyl Enter the US?

China and the Opioids

Opium is not native to China. It is native to Turkey and the region surrounding Turkey, and later imported into China. That leaves open the question of how did China, 1500 years later, become the main source of illicit fentanyl smuggled into North America?

Let's go back a couple of centuries. Rarely in history has a drug racket led directly to leadership in corporate trade and from trade leadership, directly to a war. Then, with funding from heroin smuggling, that trade leader dominated trade in its region. Asian giant *Jardine Matheson* did that and more although this is not meant to be an indictment, only to show how world trade has evolved.

Corporate giant *Jardine Matheson* was founded by Edinburgh-trained physician William Jardine, born in 1784. He became a surgeon on *British East India Company* ships that traded in the Indian Ocean waters and beyond. The *British East India Company,* founded in India in the 1600s, managed that trade. It was an enterprise leading directly to Britain's outright colonization of India that became known as the *British Raj* during their colonial rule.

Jardine teamed up with James Matheson, also a Scot, and together they became the largest importers of opium, sourced in India, and then smuggled into China. Selling opium to *Jardine Matheson*, who, in turn, sold it to China, kept the British government at arm's-length from the drug, yet still trade in and profit from opium smuggling.

Opium was introduced to China through trade with Arab merchants back in the Tang dynasty, well before *Jardine Matheson*. The Tangs ruled between 618 and 907. They were unrelated to the *Tang* orange drink on the market since the 1950s.

Opium trading with Arabs was soon supplanted by the Portuguese after Vasco de Gama embarked on his epic voyages to India and the Orient. Portuguese settled Macao in the mid-1500's and took over the opium trade entirely although Britain's hand was in it still, notwithstanding their hands off way of trading.

Opium's main use back then was as an aphrodisiac although I have no information about how effective it was for that indication. It was also touted for anxiety and pain and its use spread slowly but steadily during the next 900 years. By the 1700's, tobacco smoking spread from North America and smoking taught the Chinese a new way to take opium, because until they learned to smoke it, they took it by mouth. Pipe smoking, introduced by Europeans found its way to China before cigarettes but it didn't take long for North Carolina cigarettes to replace pipe smoking. That increased addiction and the first of several laws banning opium imports was passed. Those laws had very little effect on the spread of the abusive habit.

In his book, *1493*, [Vintage Books, 2011], writer Charles C. Mann said, referencing farmers during the Qing dynasty, "they cut their tobacco with opium." That almost sounds like the modern concept of finding many illicit drugs cut with fentanyl.

The U.S. began trading with China after the 1776 Revolution and became major importers of Chinese tea. It was during that time that Scotsman Adam Smith's theories of free trade became widely known as well, including his observation that "…competition can lead to economic prosperity."

A hundred years later, Portuguese traders continued their pattern of buying opium in India and trade it to China with a high profit margin. That also drew the U.S. into that side of international trade.

Years ago, I visited Mumbai (it was still called Bombay) and saw first-hand, the continued and widespread use of English in the region surrounding the city, a remnant of the *British Raj* colonial era. I also visited Goa, which displayed evidence of its history as Portugal's colony through the local persisting Portuguese dialect and Catholic priests walking in the street. Portugal introduced Catholicism to India and many Indian Catholics today still have Portuguese-sounding surnames, especially in the regions around Mumbai and Goa.

Britain's opium trade was coincident with a growing world-wide demand for Chinese silks, tea, and porcelain but that trade ended up in just one direction, resulting in a trade imbalance with China because China refused to import Europe's manufactured products and they tried prohibiting opium, forcing it into an even larger smuggling operation." That trade began about 1820 and continued from there. As an aside, Chinese porcelain was also how a set of our dishes became known as "a set of China."

In her carefully researched book, *The Opium War*, [Abrams Press, 2021] Julia Lovell said, "The greater part of the profits fell into the pockets of the British government, whose agents in Asia controlled opium production in Bengal. The East India Company did not publicly dirty its hands by bringing the drug to China. It commissioned and managed plantations of opium poppies across hundreds of thousands of Indian acres...Finally, it oversaw the packing of the drug into mango-wood chests, its shipping to Calcutta, and auctioning it off." She went on to say, "The missionaries became natural allies of the smugglers."

From 1644 to 1912, China was ruled by the Qing dynasty and the population was increasingly dysfunctional because of widespread and growing addiction to the very opium smuggled in from Britain to pay for the tea, porcelain and silks the Western world relished.

Britain had enough of Chinese reluctance to import anything, especially the Indian opium so they invaded China in what has become known as The First Opium War. Those hostilities were settled by the 1842 Treaty of Nanjing, between Britain and China. The U.S., a year later, signed the Treaty of Wangxia, which, among other guarantees, gave Americans the right to own land in the so-called five Treaty Ports, one of which was Shanghai, the city that led to the expression, "I was Shanghaied." Similarly, Hong Kong became a British colony as Macao went to Portugal. Quoting Julia Lovell again, "In China today, the Opium War is the traumatic inauguration of the country's modern history...but also the start of China's century of humiliation."

The Qing dynasty didn't observe the Treaties to the extent the Western powers expected them to, and in 1857, Britain attacked Chinese port cities yet again in what is called The Second Opium War. That led, a year later, to

China granting France, Russia and the U.S. treaties that guaranteed the same rights Britain guaranteed itself by attacking China. Charles Mann wrote, "British forces freely disseminated the opium that the government had gone to war to exclude."

Among the rights granted by those treaties, the island of Taiwan became an open port for Western trade. For decades, the island was known as Formosa, a name derived from Portuguese who, in 1542, described it as Beautiful Island (*Ilha Formosa*). After China adopted communism and authoritarian rule, Taiwan became known as the Republic of China while the mainland was referred to colloquially as Red China.

Jardine Matheson became a Hong Kong based corporate giant and their multiple businesses no longer include opium.

Lovell wrote, "the narcotic puritanism of twentieth-century China's two great dictators, Chiang Kai-shek and Mao Zedong -both sworn public enemies of opium, both bankrolled by drug-trade profits."

Charles C. Mann emphasized the role of drugs in Chinese politics. He wrote, "Some of the biggest producers were the British descendants of the Nationalist military officers who fled Mao Zedong's takeover of Beijing in 1949. They were joined and to some extent replaced in the 1960s by guerillas from Communist uprisings in Myanmar. Beijing was subsidizing these guerillas, its simultaneous efforts to shut down the Golden Triangle drug trade were, not surprisingly, less than successful."

More than ninety percent of the heroin in the U.S. today is smuggled in from Mexico. Most of it is heroin made from Chinese opium. In fact, in 2019, China sentenced one of its citizens to death for running a lab that produced fentanyl later smuggled into the U.S. in large quantities. Chinese authorities even cooperated with American law enforcement although they later suspended his execution and I have no follow up as to whether he was executed. Fentanyl and heroin still come in from Mexico and the DEA still says it originates in China.

Less than ten percent of the opium comes into the U.S. from Afghanistan, which, along with Turkey, is the ancestral home of the opium poppy.

"Heroin seizures almost predominantly are through the port of entry and either carried in a concealed part of a vehicle or carried by an individual," *Customs and Border Protection* Commissioner Gil Kerlikowske testified to a U.S. Congressional committee. He went on to add, "We don't get much heroin seized by Border Patrol coming through, I think just because there are a lot of risks to the smugglers and the difficulty of trying to smuggle it through."

The U.S. *DEA* reported similarly back in 2016 in its *National Drug Threat Assessment*, "Illicit drugs are smuggled into the United States in concealed compartments within passenger vehicles or commingled with legitimate goods on tractor trailers."

By contrast, the *BBC* reported last year (2021) that Afghanistan is responsible for more than 80% of the world's opium supply, but almost all of it, outside the U.S.

We must understand how heroin and fentanyl can be so easily concealed.

They're potent drugs, or, as the old commercial for *Brylcreem* hair treatment used to say, "A little dab'll do yah."

As a result of that high potency, the smuggled shipments are not very large, making them even more difficult to detect.

During my research, I found a report in a 1961 academic publication known as the *Journal of Pharmacology and Experimental Therapeutics* that allows us some rough calculations. The writers stated the clinical dose of morphine is 10 mg and they concluded a similar analgesic dose of heroin was two to four times less than morphine. "The amount of heroin needed to match the analgesic potency of morphine (10 mg) in the group comparisons ranged from 2.3 to 5.2 mg" From their data, let's say heroin is about four times more potent, so I'll set heroin's clinical dose at 2.8 mg, to facilitate my calculations. There are 28 grams in an ounce and that computes to 28,000 mg because there are 1000 mg in a gram. That makes about ten-thousandths of an ounce, or 2.8 mg, enough for one human dose. You can begin to see how small amounts multiply to dose huge numbers of users.

With those calculations, an ounce of heroin will therefor treat ten thousand people. There are 32,000 ounces of heroin in a ton of the stuff and if ten thousand people per ounce can abuse it, then 320 million people can abuse a ton. That's the entire U.S. population.

Government agencies estimate 100 tons are smuggled into the U.S. per year. A hundred tons is less than three storage containers that ride on the back of semi-trailers, the entire amount of heroin smuggled into the U.S. annually. That's why smuggled amounts are small, because of heroin's potency.

That also makes heroin's small dose a big problem, both domestically and, with the drug coming out of Afghanistan headed for the rest of the world, internationally too.

It's easy to hide.

The small package size of smuggled heroin points to another problem. Many prescriptions in the U.S. are filled by mail. If we think about that, tablets or capsules are in a closed plastic container, perhaps with the container enclosed in a padded envelope. The United States Postal Service can inspect only a small fraction of the mail so most opioid prescriptions in the mail are delivered to the recipient. I think mailing opioids should be an activity of the past and not of the future, at least until we have a mechanism for inspecting all mail with a technology that I've labeled "sniffer" technology. It must be able to "sniff" tens of thousands of packages daily.

Heroin's problem is magnified when fentanyl is mixed with it.

Fentanyl is 100 times more potent than morphine, so if morphine's dose, as I said above, is 10 mg, fentanyl's dose is 1/100th that, or 0.1 mg. That's a tenth of a mg or 100 micrograms. A microgram is a millionth of a gram. Not much stuff.

According to the U.S. *Drug Enforcement Administration*, or *DEA* (https://www.dea.gov/resources/facts-about-fentanyl), "Illicit fentanyl, primarily manufactured in foreign clandestine labs and smuggled into the United States through Mexico, is being distributed across the country and sold on the illegal drug market. Separate studies show that the fentanyl smuggled in from Mexico, comes from China, along with the opium. Fentanyl is mixed in with other illicit drugs to increase the potency of the drug and the product is sold as powders and nasal sprays, and increasingly pressed into counterfeit pills made to look like legitimate prescription opioids."

A separate, unclassified DEA Intelligence Report, *Fentanyl Flow to the United States*, reported. "Effective May 1, 2019, China officially controlled all forms of fentanyl as a class of drugs. The implementation of the new measure includes investigations of known fentanyl manufacturing areas, stricter controls of internet sites advertising fentanyl, stricter enforcement of shipping regulations, and the creations of special teams to investigate leads on fentanyl trafficking. These new restrictions have the potential to severely limit fentanyl production and trafficking from China. This could alter China's position as a supplier to both the United States and Mexico."

That unclassified intelligence report went on to say, "In 2017, the DEA provided information to India's Directorate of Revenue Intelligence, resulting in the takedown of an illicit fentanyl laboratory in Indore, India in 2018."

Despite China's crackdown on fentanyl's production, and India's competition with it, the DEA concluded, "The flow of fentanyl to the United States in the near future will probably continue to be diversified." Diversification in this case means smuggled in small batches hidden in either cars or trucks, or already mixed into other drugs.

Illicit fentanyl manufacture has an additional issue. How does a lab maintain enough precision in its measurements to ensure humans are safe taking the drug? That's especially so considering fentanyl's potency.

When I managed my own lab, I had a $2000 electronic scale that could safely and reliably measure 1.0 mg. Clandestine labs don't have such a scale, so their quality control and reliability are missing. A 1.0 mg dose of fentanyl could be off by a factor of two and 2.0 mg could be a lethal dose in a small person.

As I said, fentanyl can magnify heroin's problem, especially, when it's mixed with heroin to increase heroin's potency, allowing the drug dealer to get more mileage, if you will, from his supply of heroin. Because the amount of fentanyl the dealer adds could be a lethal dose.

As the DEA says, "One kilogram of fentanyl has the potential to kill 500,000 people." One kilogram is two and a half pounds. Not much stuff, and easy to smuggle as part of a tractor-trailer load or shoved under the dashboard of a car.

The DEA reports further, "Drug trafficking organizations typically distribute fentanyl by the kilogram (1000 grams). As we learned in the previous paragraph, one kilogram of fentanyl has the potential to kill 500,000 people."

The DEA went on to quote the Center for Disease Control, or CDC, "According to the CDC, synthetic opioids (like fentanyl) are the primary driver of overdose deaths in the United States. Comparison between the 12 months-ending January 31, 2020, and the 12 months-ending January 31, 2021, during this period, overdose deaths involving opioids rose 38.1 %," to the staggering total of 100,000 needless deaths. In 2022, the total continued to rise and hovers at 108,000 needless deaths. It's a death rate that shows no signs of abating and for a decade it has continued to rise every year.

Chinese fentanyl is not only incorporated as a contaminant in heroin, but factories are also manufacturing copies of licensed prescription fentanyl tablets and sending those abroad too.

There are recent news reports that fentanyl is mixed into other drugs so frequently, it is being blamed for the bulk of overdose deaths in the U.S. The source of fentanyl is indeed diversified, users can buy it through social media, like *Facebook* and *Instagram*.

That's ironic because social media sites popularity grew as a substitute for the decreased social contact Americans began to display. Social media allowed us to establish and maintain social contact with strangers, albeit electronically. The loss of social contact is blamed for increasing opioid use, suggesting social media should mitigate drug use. Yet social media has become the new drug dealer, entirely opposite from our expectations and that modality is blamed for increasing the overdose death rate in young people.

It is at once ironic and paradoxical that so much effort to conquer fentanyl distribution can be expended while at the same time, executives managing legal fentanyl sales, marketed by a drug company as *Subsys* and introduced to treat cancer pain, could be punished for excessive promotion.

The company is Arizona-based *Insys*, one of whose founders, John Kapoor was ordered to prison in 2020, for five and a half years, for too aggressively marketing the under-your-tongue fentanyl spray. Other Insys officers were sentenced to shorter prison terms.

It's paradoxical because while the U.S. lobbied internationally for reduced fentanyl imports, it was still being hawked legally, although far too aggressively, by American companies. The result was people without cancer were prescribed *Subsys*.

Good Reception

Pharmacologists always teach students that drugs are defined as chemicals that bind to our body's receptors, and by binding to those receptors, they trigger a response by the cells that have those specific receptors. The argument is circular because if we ask, "Where are those receptors?"

The answer always comes back as, "In your body," or, "in your brain." Not very specific. We learned the right answer indirectly.

I'll use my own family to illustrate. My older son is an insulin-dependent diabetic, sometimes called a Type 1 diabetic. While his condition doesn't relate to opioid use, uncovering insulin's mechanism of action led us to the physical discovery of what we call our receptors including our opiate receptors.

It's a late twentieth-century story.

Insulin was discovered in the 1920s by Frederick W. Banting and his technician Charles H. Best in a Canadian lab. Banting shared the Nobel Prize for his discovery and gave his patent rights to *Eli Lily & Co* because he wanted insulin to be available generally. It's an example of altruism of which we don't see enough. Separately, insulin was produced in Europe by *Novo Nordisk* and by the mid 1920s it became widely available in the Americas and Europe.

Half a century later, in 1970, labs around the world showed it worked by binding to cell membranes and that act of binding signaled our cells to take up the sugars circulating in our blood. That's how insulin lowers our blood sugar, it binds to what pharmacologists always called receptors, specifically, insulin receptors. All of pharmacology had been built on presumed drug-receptor interaction. The Insulin Receptor was the first to be shown to be a physical location on the outside surface of a cell membrane. Cell membranes are the enclosures that wrap around cells and keep their contents intact.

The insulin story has more built into it because insulin is a hormone we make in our pancreas. It is also a drug we can buy at the drug store, formulated so diabetics can inject it. So, it makes sense retrospectively that we have insulin receptors because we make insulin. Insulin-dependent diabetics don't make enough insulin, so their insulin receptors lie empty, unable to tell cells to take up circulating sugar. We can ask, "Why would we evolve insulin receptors?" We know the response now is, "Because we make insulin, and our bodies needed a mechanism to use it, so we also evolved insulin receptors."

I also learned about opiate receptors as a graduate student because receptor theory is such a basic pharmacological principle. As I said, all drugs work by binding to receptors. Then in 1973, Candace B. Pert, herself a graduate student at Johns Hopkins University, together with her mentor, the eminent Solomon H. Snyder published a scientific paper entitled, *Opiate Receptor: Demonstration in Nervous Tissue*. I was a third-year graduate student by then at Boston University, in the Department of Pharmacology at the medical school, and like every other graduate student of pharmacology around the world, marveled at her discovery. Our department meetings began to revolve around discussions of opiate receptors.

Her discovery, however, also opened the question that if we have opiate receptors, what are they doing there? Do our brains make something that binds to those receptors? Opiate research around the world began to search for an opioid that our brains make that bound to those receptors. Analogous to insulin, nature, evolution, or a higher power wouldn't have put opiate receptors in our brains if our brains didn't make something that uses them.

Adding to that conundrum, our white blood cells, known as leukocytes, also have opiate receptors on the surface of their cell membranes. Does that suggest opiate receptors play a role in our immune response.

Yes.

Discovery and demonstration of the insulin receptor stimulated, so to speak, other labs around the world to look for opiate binding to cell membranes in part because the mathematical and graphical representation of opioid drug binding to its theoretical receptors, was identical to how insulin could be shown to bind to its insulin receptors so there was some basis for the idea that there was a physical opiate receptor we could see. The thinking was, "Maybe opioid drugs worked like insulin worked," specifically by binding to cell membrane receptors to give our nervous systems the message, "It doesn't hurt anymore."

But opiate receptors were different because when scientists discovered that we had measurable opiate receptors, as far as anyone knew, we didn't make morphine or any other opioid. That left us metaphorically scratching our collective scientific heads and saying, "If we don't have circulating opioid hormones, why would opioid receptors have evolved?"

Turns out that we do have opioid hormones. Scientists in the U.S. and Britain figured out how morphine and those opioid drugs work back in the mid-70's while I was still a graduate student. The first hint to its mechanism of action was when several labs reported in the scientific literature that morphine and other opioids "bound" to brain tissue. That needs its own explanation because it goes back to the very definition of a drug.

The invention that made the definition come alive by modeling drugs binding to receptors was when drugs were made radioactive and gained the nickname "tagged" molecules. That technology came out of the nuclear industry and meant the drugs were manufactured with a radioactive tracer like tritium in place of a hydrogen, or a radioactive carbon, known as carbon-14, in place of a plain old carbon atom. The point is those "tags" are mildly radioactive, by emitting electrons that can be counted by a special lab instrument. It's as though nuclear physics married biology and their offspring made us say, "Oh, I can see the drugs bind and there really are receptors."

Those early investigators found that radio-labeled or tagged opioids and natural opiates stuck to brain tissue, not just stuck the way ink stuck to a towel, they stuck reversibly, what scientists call 'binding, that is, they were bound to their receptors. The same opiate drugs without radiolabels could push the radio-active opioid from its binding site in brain tissue. The more unlabeled drug added to the brain tissue, the less tagged drug was bound there. Those binding places were real drug receptors and the whole mechanism satisfied the analogy we were taught that drug receptors are locks and drugs are keys to the locks. Opiate receptors were, like insulin receptors, evolved because we have opiate chemicals in our brains just as we have insulin in our pancreas. The opiate chemicals needed a place to work and, like insulin has its receptors, so do the opiate chemicals. Pharmacology was changing. We could now measure those receptors.

Pharmacologists were always taught, "Drug receptors are specific areas of our cells to which the chemicals in our bodies, stick to produce their effects and regulate our metabolism." Chemicals like hormones and neurotransmitters are members of that chemical signaling system. Our local area network, to borrow a concept from the Internet.

Those investigators thought about receptor binding and concluded, "If a drug we add, binds to the tissue reversibly, there must be something our bodies make that binds to the same receptor." They were thinking why we would evolve opiate receptors unless our bodies made an opiate that uses those receptors?" Unlike insulin, that was a hormone in search of a receptor, opioid drug receptors turned the idea around. They were receptors in search of a hormone.

That thinking triggered a search for circulating opioids. Initially, it seemed far-fetched because we don't make willow bark or poppy seeds, but in the mid-1970s, chemicals with opiate activity were discovered in brain tissue and to everyone's surprise, they were peptides, strings of amino acids in a programmed sequence. That sequence of amino acids meant our bodies made them according to instructions from our DNA. We not only evolved them for a purpose, but we also inherited the sequence from our parents much like our hairline, height, weight, and shape of our noses!

That discovery opened the question, "What purpose could our own opiates serve, and did they tie into pain management, drug abuse or addiction?" Adding to those open questions, is why scientists could differentiate three types of opiate receptors, known by their Greek designations, Mu, Delta, and Kappa. Sounds like a fraternity. Labs have even cloned the genes that code for those peptides.

Those questions have occupied us ever since. We now know our bodies use those peptides to manage our reactions to stress and pain. The activity that was unpredictable is the lack of "opiate" activity those peptides show and that we don't become dependent them although there is a theory that gym memberships and workout schedules are driven by exercise-induced feeling of well-being, because exercise raises levels of our endogenous (circulating) opioid peptides. The opposite of endogenous is exogenous which means swallowed, injected, inhaled, or rubbed on to our skin.

There are results of some amazing experiments. Scientists have, through genetic engineering, raised mice missing one or the other of the three types of opiate receptors. They reported experimental results showing when mice were raised missing their Mu receptors, morphine's analgesic activity in those mice was abolished. They concluded that Mu opiate receptors are molecular switches that reinforce their own stimulation. More simply, morphine binds those Mu receptors and that makes us feel good, so we take it again.

Mice raised without their Delta receptors show increased anxiety levels and suggest drugs that bind to our Delta receptors would make good anti-anxiety treatments.

Now that I've said Mu, Delta and Kappa binding sites are receptors in search of a hormone, what are the opiate hormones?

They are known as Dynorphins, Endorphin and the Enkephalins and all three are released by the nerve cells in our brains.

What's their function?

They bind selectively to our Mu, Delta and Kappa opiate receptors and the circulating opioid peptides and their receptors form a physiological mechanism that helps us manage anxiety and pain.

Their release by our brains also depends on our internal stress levels. Early experiments with naloxone in human volunteers showed stress activates our endorphin pain-suppressing system. Stressful pain was made worse by naloxone showing that stress activates our endorphins, and our brains secrete more of it. Endorphins are our own anti-anxiety system.

I don't want to get too far ahead of my story, but those three opiate receptors suggest a strategy to design non-addicting analgesics. They could have the analgesic strength of morphine without morphine's addictive potential.

Sublimaze, Philly Dope and China White

I'm a pharmacologist, albeit retired, and I write about drugs, not toys. A few years ago, however, my now-thirteen-year-old son received a gift with the name *Smithsonian* prominently displayed on a shrink-wrapped package of four boxes and I automatically assumed it was an educational toy because of the museum name. I picked it up to read the directions, but a *Made in China* imprint stopped me. Not only does the Smithsonian sell toys from China, but I can't find a toy in my house, other than *Legos*, that wasn't made in China.

This is a book about opioid drugs though, and even the illegal drug market, especially fentanyl, is changing as the drug trade also seems to be relocating to be cornered in the Far East, mainly in China. An article I read online in *Bloomberg* written by Esme E Deprez, Li Hui and Ken Mills reported, *Deadly Chinese Fentanyl is Creating a New Era of Drug Kingpins*. Similarly, one written by Edward Helmore published in *The Guardian* (Dec. 27, 2017) entitled, *It's all fentanyl: opioid crisis takes shape in Philadelphia as overdoses surged*.

So, both our toys and our street fentanyl are now Made in China suggesting the drug trade parallels the market in other manufactured commodities. Although heroin still comes from Afghanistan, India, or Mexico, most street fentanyl can be traced back to China. Of course, fentanyl is smuggled into the U.S. through Mexico and more and more frequently appears as a "contaminant" of heroin and other drugs. Fentanyl's street names reflect its origin, *China White* or *China Girl*, and like toys and tools, fentanyl smuggling has emerged as a global trade, yet still funneled through Mexico. Fentanyl derivatives mixed with heroin are appearing on the streets and the fentanyl mixed with other drugs is dangerous and responsible for more deaths from overdose. For example, one of those derivatives is called carfentanil and it was synthesized as a large animal tranquilizer, usually administered by a tranquilizing dart. Large animal refers to bears, elephants, and horses. If it was designed to tranquilize a bear or a horse, it's no wonder that it will kill people.

I've shown morphine and codeine are the two well-known opiate products from opium poppies and they have long histories. Fentanyl is the opposite because none of it comes from plants. It is a twentieth-century invention, entirely synthetic. Yet it is still a very powerful morphine-like drug that originated at a drug company.

Dr. Paul A.J. Janssen, a talented Belgian chemist, founded a small drug company in 1953 in Beerse, Belgium and named it *Janssen Pharmaceutica* after himself and, he claimed, after his father. He synthesized fentanyl in his laboratory in 1960. His starting material for that new opioid was the drug meperidine (pethidine in Europe), known more commonly by its tradename *Demerol* which itself was invented and patented in Germany, back in 1937, by *I.G. Farben* and reached the market in in 1943, right in the middle of the Second World War. I discuss that more in a separate chapter, but he began with a synthetic starting material, not with a plant product as heroin which we've seen is just a simple derivative of morphine.

Before Paul Janssen could launch his new product fentanyl, in 1961 he sold his company to *Johnson and Johnson* (*J&J*) and it became a wholly-owned subsidiary of *J&J*. We can only guess at J&J's motivation, but we must give them some credit in recognizing fentanyl's market potential. *Janssen* launched fentanyl in Europe in 1963 but *J&J* had to wait five more years for FDA approval before they could launch it in the U.S. They launched it as *Sublimaze* in 1968 and marketed it as an anesthetic for surgery. It would dominate the general anesthetic market for the next thirty years. I received it twice in a week for "procedures" during a recent hospital stay as I worked on this book.

As I began this piece, I was lying in that hospital bed waiting for that "procedure," hospital parlance for something that would hurt. I received a coronary stent, like a ballpoint pen spring, but my spring lets blood flow unimpeded to my heart. I was forced to lie still for three hours and not eat anything after midnight, which invariably left out breakfast. A resident explained, "After they numb you, they'll give you fentanyl to relax you. Do you have any questions?" she asked, eyebrows raised, as though I should have been pleased, or grateful for her explanation.

Fentanyl probably relaxed me, but I'll never know because the drug put me right to sleep and I awakened two hours later. I knew how long I had slept because my cellphone rested on my chest through the entire procedure, and I looked at the time before I fell asleep and as soon as I awakened. I assume I was relaxed too, although not relaxed enough to drop my phone. At least I awakened in a good mood, but when I tried to see what the surgeons were doing by looking at the wide-screen TV over my "bed," I heard in firm, almost angry tones, "Please stop moving."

Those were my body parts on that TV, and I wanted to see what the doctor was doing. I guess it was their hospital though and he wanted me to lie still.

I'm fine though; I still split my own fireplace wood… more slowly than I used to but what's the rush?

Why did fentanyl's medical use spread so rapidly? The answer begins with appreciating that fentanyl distinguishes itself from morphine by being as much as a 100-times more potent than the natural product. A tiny dose is all you need, shown by a quick, easy calculation. A "normal" human dose of morphine is 10 mg or a hundredth of a gram (there are 28 grams in an ounce, or 28,000 mg, so it's not much stuff). A corresponding dose of fentanyl would be as little as 0.1 mg. We read that as a tenth of a mg or 100 micrograms and we can also say that is one-ten-thousandth of a gram. Either way, we do not need very much fentanyl to produce some very dramatic effects. There are even fentanyl derivatives, like sufentanil, in addition to carfentanil, with active doses one-thousandth that of fentanyl itself. So much more potent that either one can only be used only as an elephant tranquilizer because it isn't practical to measure a billionth of a gram reliably to make it safe to use in humans.

Fentanyl's other advantage is its two-hour duration of action, compared to morphine's, the label of which says should be repeated every four to six hours. Fentanyl is both short-acting and very potent.

Those two features grew fentanyl's clinical market share, especially its use in hospitals as I showed it was used on me...and I lived to write about it.

The pharmaceutical industry also grew fentanyl's market share by offering new formulations. A formulation to a drug company is like a new sauce to a food company. Food companies can market a chicken breast in a red sauce, white sauce, in olive oil, or breaded and they call it extending their line. A drug company can market a drug as a tablet, capsule, lozenge or in a device that sprays it up our noses. They sell creams, patches, and skin sprays, among others. Those new formulations offer marketing advantages including but not limited to, changing a drug's duration of action, altering its dosage schedule from twice a day to once a day or changing the route of administration, for example from a drug that must be injected to a drug that could be taken orally as a capsule or tablet. Each of those new formulations can be patented with the same tradename because tradenames are trademarked and not only can they be renewed, but a company can build a marketing program on a family of formulations with a single tradename. Those ploys give the drug companies marketing strength.

In the early 1980s, a small company in Mountain View, California, the *Alza* Corporation developed a fentanyl patch they marketed as *Duragesic*. It was only the second drug patch they developed as *Alza* became well-known for its drug delivery technology. *J&J* bought *Alza* too, demonstrating the trend of companies to grow by merging. *Alza*'s first patch was to prevent motion sickness. A patch allows patients to take a drug continuously without pills, capsules, injections, or the need to remember what time to take the next dose. Users must answer the question, "Is that a Band-Aid?" The patch releases its drug to our skin, and we absorb it across our skin.

The name *Alza* is an acronym of the first two letters of the name Dr. Alejandro Zaffaroni, the brilliant man who not only founded *Alza*, but five other companies as well, sequentially. He even had a role in the first birth control pill.

When I managed drug development for a pharmaceutical division of Germany's *BASF*, one of my projects was partnered with *Alza* and I flew to California monthly to meet with them. My office was in New Jersey and during one trip, I ended up in their offices on a Friday afternoon. I recall that suddenly, work stopped entirely as Alza's president appeared, introduced himself and said, "You're invited to our TGIF party." He handed me an *Alza* sweatshirt, and someone told me they partied like that every Friday. All I did on Friday's, was sit on an airplane and watch a movie although I still have my *Alza* sweatshirt, folded neatly in my closet next to my Boston University sweatshirt, my professional history tracked by a cotton blend.

Fentanyl's use expanded further in 1984, when *Janssen* incorporated fentanyl into a lollipop-like delivery system they named *Oralet* and later renamed it *Actiq*. It was even colored red and sweetened although it took nine years before FDA approved it for public use.

Fentanyl's use as a general anesthetic changed anesthesia practices because doctors found it easier to use than existing products. All they had available up until fentanyl were the so-called volatile anesthetics of which chloroform was the first and oldest example although since abandoned because of built-in side effects. They also had morphine, but that was easier to use to treat pain than as a surgical anesthetic.

Nitrous oxide was around for many years and picked up the name "laughing gas," bestowed on it by its discoverer (Sir) Humphry Davy, who experimented on himself back in 1800. It was introduced into medicine in 1844 and, although it has some analgesic properties, by itself, nitrous oxide doesn't produce anesthesia. It must be combined with something else. Davy became famous because he went on to discover electricity. His is one of the names scientists add to our vocabularies during our long educations.

Beginning in 1846, ether replaced chloroform and it was first administered by a dentist, Dr. William T. G. Morton, for surgery in what is maintained today as the historic *Ether Dome* in the old wing of Boston's *Massachusetts General Hospital*. I toured the facility way back in Graduate School and I should have known then that my interest in the history of medicine is stronger than my interest and ability to discover new ones. It's never too late! Besides, I don't want to travel for business anymore.

Halothane replaced ether in 1956 and halothane's popularity grew rapidly because all the volatile anesthetics up until halothane was invented all had problems.

Chloroform, as I mentioned, turned out to be toxic, ether is flammable and can explode and nitrous oxide is an incomplete anesthetic although it is still used as a rocket propellant or to inject into car engines for drag racing. Nitrous oxide's worst feature is that patients don't fall asleep with that gas when it's given alone.

The anesthetic agents used today are all volatile, that is they must be inhaled when they are used. They are more modern and include *Ethrane, Forane, Suprane*, among others. None of them will explode or poison us but they are used when the scheduled surgery is expected to be long. Aside from a few side effects, those new volatile agents have one advantage, if the anesthesiologist turns off the machine that administers the anesthetic, the patient awakens within a few minutes. Because these anesthetics are given during artificial respiration during surgery, the anesthesiologist can also increase or decrease the dose minute by minute if (s)he sees physiological changes that need intervention.

Intravenous fentanyl needs less setup time with fewer equipment needs than volatile anesthetics and anesthesia was its main use for decades.

But fentanyl's use also changed from a useful, short-term anesthetic, to an illegal drug of abuse as users and dealers discovered how potent it is. Smuggling a five-pound bag of fentanyl works out to several thousand dollars of illicit drug. That's backed up by statistics. In a 2017 *Journal of the American Medical Association* (*JAMA*) article, Senior Writer Rita Rubin reported that in more than half, 56%, of the opioid overdose deaths in 2016 in the ten states that make up the Center for Disease Control (CDC) monitoring program, they could detect fentanyl.

It also moved out of the cities into the country, and it is now mailed directly from China or, as I mentioned, sent to Mexico from China and then smuggled into the U.S. from Mexico. Some writers discuss how fentanyl's use was responsible for the spread of opioid abuse from inner-city ghettoes to middle-class suburbs, but sociology is not my strong suit.

The *Bloomberg* article I mentioned earlier in this chapter established how and why the street fentanyl came in from China. There are other reports too about that niche of international trade. *TheGuardian.com* reported on Jan 24, 2018, *Chinese Labs Use Mail to Send Opioid Fentanyl into U.S. Senate Report Finds.* That implicates the U.S. Post Office although not because they are at fault or in any way culpable, but because they are ill-equipped to handle the influx of illegal chemicals shipped in plain brown wrappers. There is a good opportunity for technology to be harnessed in drug detection. Even my car beeps at me when I am approaching another parked car and I'm certain technology can be designed to detect fentanyl in a plain brown wrapper at the Post Office or hidden in a car or buried in a truckload of goods.

Similarly, *latimes.com* reported on Dec. 13, 2017, *Smuggler Busted With Almost 80 Pounds of Fentanyl at U.S. Mexico Border, Agents Say.* That's further support for the reports that Chinese fentanyl is smuggled into the U.S. across our border with Mexico, as opposed to coming in directly from Asia.

My interest is how do we stop it because it presents us with a drug abuse problem that is different from any we have had in at least two generations where heroin, devoid of medical use, was the drug of choice. Fentanyl, sold as *Sublimaze* is still a legal, useful, and safe drug used widely in clinical practice. The *Duragesic* patch has a respectable market in treating severe chronic pain such as caused by cancer, a market also addressed by *Subsys*, the sub-lingual formulation sold by *INSYS*. The fentanyl lollipop *Actiq*, although now restricted to use in children over 16 years old because it was abused in the past, also treats chronic pain. That's the legal side, although we'll see shortly that the legal side and illegal side collide sociologically.

On the illegal side, fentanyl mixes well with other drugs and one of the most common mixtures is *Philly Dope*, which is heroin mixed with fentanyl. That mixture resulted when dealers cut heroin so they would have more product to sell, but cutting it decreased heroin's potency, so the dealers countered that loss of potency by adding a little fentanyl. Remember, fentanyl is 50 times more potent than the heroin and 100 times more potent than morphine, so it raised the cut heroin's potency. The problem that strategy created is that it began to kill users. *The Guardian* article I mentioned above reports a 540% increase in fentanyl deaths between 2014 and 2017.

Fentanyl's incredible potency also means smuggling requires only small packages, think of a 5-pound bag of sugar riding in a semi-trailer; how does an inspector look for something that small hitching a ride in something so big? Fentanyl shipments even arrive by FedEx and UPS. I repeat my wish, that someone will design a "sniffer" technology that could "sniff" our shipments hidden on semitrailers or in a hollow hiding place in a passenger car.

Fentanyl presents the U.S. with a complex commercial problem. It is a clinically useful prescription drug that has spawned a sophisticated international smuggling operation.

I don't have data, but I'm certain most people will agree that toys, tools, cooking gear and many other manufactured products are all made in China and imported into the U.S. because the profit margins are higher than they are for products that used to be manufactured in the U.S. All our factories moved to China twenty-five years ago to save money as our labor costs increased. The only visible result is big factory spaces have begun appearing as gallery space rentals, loft apartments that feature high ceilings or large, vacant shells.

The Chinese have promised to try to solve the fentanyl problem and Congress is in on the case too but simply outlawing it and increasing the penalties will not solve the issue because the market we present to Chinese fentanyl dealers is too good and the U.S. mail service is not equipped to intercept all the fentanyl that has begun arriving through the mail or the drug shipped through Mexico and smuggled into the U.S. on the highways.

We are not stuck with it. I believe there is a way out, but we have to change our strategy.

I used to think that if the U.S. said to China, "We'll work with you to get control of fentanyl manufacturing in your country and the illegal export to the U.S. But, if we don't begin to see a trend toward less illegal Chinese fentanyl circulating in the U.S., the import duty on Chinese toys will begin rise to cover our law enforcement costs. If that doesn't solve the problem, import duty on other goods manufactured in China will also rise." The problem with that strategy is that our current administration has raised duties and all it has accomplished is threats from Asian countries to cut back on exports to the U.S.

I think the only way Asian countries would get the message is if their own citizens were increasingly plagued by addiction to opiates. That's not something we can influence. I can't even find statistics for Asian countries.

As I said, we must fall back on technology. Can we do a better job intercepting shipments? Can we catch smugglers and exclude them if they are from other countries, or jail them if they're American? Can we work with shipping companies to inspect trucks more thoroughly and identify illegal shipments at FedEx, UPS and the US Post Office?

Technology allows us to open our doors, turn up our thermostats, start our cars, listen to country music through wireless earphones and turn on our lights. It's not too big a stretch to imagine depending on technology to find a bag of fentanyl on a semi-trailer arriving from Mexico loaded with vehicles or agricultural products. That part of the market is known as *The Internet of Things*.

Let's create the *IDS*, the *Internet of Drug Smuggling*.

We know smuggling is illegal yet the latest headlines about fentanyl bring it much closer to home as a prescription medicine. A January 23, 2020 *NY Times* article written by Katie Thomas was headlined, *Insys Founder Gets 5 ½ Years in Prison in Opioid Kickback Scheme*.

Wait a minute, domestic fentanyl, manufactured by a drug company and sold by prescription, resulted in the drug company's founder being imprisoned. What's wrong with this picture?

A fellow named John Kapoor founded *INSYS* back in 1990. He was born in India and made his fortune by taking another pharmaceutical company public. *INSYS* began marketing *SUBSOS*, which is fentanyl formulated to be sprayed under your tongue. Its onset is rapid, and no needles are involved, and it was marketed for severe cancer pain.

But seven *INSYS* executives were accused of violating statutes related to drug advertising, marketing and they were even accused of bribing physicians to prescribe the sublingual fentanyl formulation. All seven executives were sentenced to prison terms.

Fentanyl is so attractive to drug smugglers and the pharmaceutical industry that now both smugglers and company executives can meet...in the prison yard.

The Big D

Longer ago than I care to admit, when I was a freshman at the University of Cincinnati, I spent the first six months behaving as though I was still a teenager in Cranford (NJ) High School. Most of the other "dorm rats" did too and I was horsing around late one evening with everyone else who lived on the second floor and part of our horsing around made me physically block a fellow, first name Bob, from leaving his room. I was having fun and without thinking, I had even hooked my thumbs around his doorjamb so he couldn't push me out of the way. His response made as much sense as my action. He closed his door forcefully and even locked it.

When he slammed it, I couldn't move my right hand out of the way fast enough and my thumb was crushed in the door jamb.

As blood encircled my wrist and began to run down my arm I hollered, "Open the door my thumb's crushed."

"I don't believe you," came the reply from behind the door.

I screamed again, "Open the door, it hurts."

Someone else also yelled, "Bob, open the door, he's really hurt." A freshman crowd gathered very quickly.

"Ooh, look, he's bleeding," was repeated several times.

Bob opened the door, and I pulled my hand away to look at it. The end of my thumb was hanging off, connected to the rest of the thumb just by a piece of skin. The force popped off my thumbnail which I put into my pocket and saved as a trophy for few years. I grabbed my thumb and approximated where it belonged as a dorm-mate called the campus police who brought me to the nearest hospital emergency room. An E.R. surgeon who had already worked all night injected a local anesthetic and made a first attempt to reattach my thumb. It hurt so much that I shouted, "Ow, that hurts!" I squirmed and stared at him.

"Give him the big D," the surgeon ordered.

A nurse injected me, and the pain went away. The Big D was *Demerol* and one of its side effects is that it causes hallucinations.

To this day, I have a vivid memory of watching myself fly through the hospital corridors with my arms spread like wings as I waited for the campus police to take me back to my dorm. I was hallucinating and the vision has stayed with me all these years. Too bad I can't draw.

That story of my youthful foolishness demonstrates two characteristics of Demerol, (1) it is a narcotic analgesic useful to treat acute pain, especially after surgery, and (2) it has side effects, including but not limited to hallucinations.

It is also widely abused on the street where it is still called "D," but also called Demmies or Dust.

Demerol is the trade name for the chemical known as meperidine in the U.S. and pethidine in Europe. It is a synthetic narcotic, and I don't know how it got two generic names, but it works the same way morphine does, by binding to the opiate receptors in our brains. That *Demerol* characteristic wasn't predicted when German chemists synthesized it way back before World War 2. Chemically, its structure doesn't resemble morphine, yet its mechanism is the same, one of the mysteries of chemistry that is exploited by chemists the world over. They can synthesize chemicals that have the same actions as a natural product. Morphine is a natural product, but a drug company can patent the synthetic chemical, they can't patent the natural product.

Demerol's activity differs from morphine's only by its dose and duration of action. It is just one-tenth as potent as morphine, so if morphine's clinical dose is 10 mg, Demerol's is 100 mg. It is also shorter acting than morphine, so we not only have to take more of it, but we also have to take it more frequently to make whatever is hurting us, stop hurting. That theme of taking away a bad feeling rather than creating a new one, we will revisit frequently as we discuss opioid addiction.

Unfortunately, when we need to get rid of pain, we also set ourselves up for drug-seeking behaviors. That doesn't make sense until we think about it.

Demerol is not only shorter acting and requires a higher dose than morphine, but we also quickly become tolerant to it. Tolerance is the natural tendency of some drugs to seem less effective with each dose. That makes us want to take more of a drug and take it more frequently than the dosage instruction that accompanies our prescription. We feel better when we take it and begin feeling worse as it wears off, so we want to take it again, even if that feeling of wanting to take it again is hours before the instructions tell us we should take the next dose.

I was surprised to learn that *Demerol* was synthesized way back in 1938 by a chemist at *I.G. Farben*, the chemical conglomerate assembled by the Nazi government, to support the regime by decreasing its dependence on imports. After World War 2, *Farben* was broken up and coincidentally, *BASF* (*Badische Anilin und Soda Fabrik*) emerged from the corporate wreckage. They were my final employer before I retired. and I enjoyed my three years with them although my required travel to Ludwigshafen, Germany and Nottingham, England every other week took its toll on my health. Today *BASF* is the largest chemical producer in the world, and they got out of the pharmaceutical business by selling the division that employed me.

Two days after my dorm accident, my thumb became infected and it hurt more and more so I found an orthopedic surgeon in Cincinnati who said, "Meet me at the hospital tomorrow at 10 AM and I'll rebuild your thumb."

"As an outpatient?" I asked.

"It'll take about an hour."

The next morning, Bob, the fellow who closed the door on my thumb, lent me his car so I could drive to the hospital. Someone led me to a small operating room where my orthopedic surgeon was washing his hands. "Have a seat," he said, "put your hand on that towel and relax. This will be over before you know it."

He injected my hand with what I assume was a local anesthetic, took an x-ray and went to work. He was manipulating pieces of my thumb bone with a pair of tweezers. "You did a job on this thumb," he said as I was forced to look away. I didn't think it was a good idea to look at my own bones.

An hour later, he said, "You're all set."

I drove back to my dorm. Two days later it was obvious that I couldn't write, and final exams were a few weeks away. I called my folks and told them about my experience. "The worst news of all is that I can't take finals because I can't write so I'm going to stay here and register for the summer for the two courses in which I'll get incompletes."

It was the first time, but wouldn't be the last, that I didn't stay with my parents during my summer vacation. It was a brutal, but appropriate introduction to the independence that accompanies adulthood.

Demerol is not the oldest synthetic or semi-synthetic opioid nor is it the most widely abused. But it is the only synthetic opioid with which I have first-hand experience, so to speak.

It was also the first and last time I am aware of that I received *Demerol* because I would have recognized the hallucination.

A Lifeguard

How many of us have ever seen the sign posted poolside, *Pool Closed. No Lifeguard on Duty*?

How many of us are beginning to see naloxone available in some unusual places such as vending machines, police stations and ambulances? Naloxone is the new lifeguard although it had a precursor.

Back in 1954, a drug with the tradename *Nalline* was introduced as an antidote to opioid overdose by our old friends at the American company *Merck*. Chemically, it was known as N-allylmorphine and, while nearly unpronounceable, we can see 'morphine' embedded in the name suggesting it is a chemical derivative of morphine.

While morphine is the classic agonist, naloxone is only an antagonist of opioids. *Nalline*, paradoxically, had both activities, it was a mixed agonist and antagonist.

How could that be? I mentioned there are three opioid receptors, *mu*, *delta* and *kappa*. Nalline is an antagonist at the *mu* opioid receptor, where morphine acts, and an agonist at the *kappa* opioid receptor. By stimulating the *kappa* receptor, it produced hallucinations, anxiety, and confusion. What's more, it's effect on the patients changed according to whether they had taken an opioid or not.

As far back as 1915, *Nalline* was shown to prevent or abolish morphine-induced respiratory depression, the lethal side effect, but it wasn't really studied until 1950 when it was shown that in animals or patients who hadn't received an opioid, *Nalline*'s effects were mild and not unpleasant. The effects resemble morphine's effects by decreasing our heartrate, lowering our body temperature and making us sweat.

But if the patients had taken morphine or other opioid, Nalline's effect differed from morphine in that *Nalline* completely and rapidly reversed morphine's effects in patients who had taken it. If the patients were addicted to an opioid, *Nalline* produced a withdrawal reaction. As I said, it was an antagonist and an agonist, depending on what the patients had taken.

That antagonist effect was exploited when *Nalline* was used back in the fifties. It was a tool used to diagnose opioid addiction. If patients given *Nalline*, showed withdrawal reactions, they were diagnosed as addicts. The withdrawal reaction was so dramatic, Nalline ended up in court as "Cruel and unusual punishment" and using it to diagnose opioid addiction was discontinued. Today we have simple blood tests to determine whether patients have an opioid in their bodies.

Naloxone, by contrast was licensed by FDA in 1971 although its chemistry was originally studied in a different environment, part of a cancer drug program at what was *Sloan Kettering* in New York City. It was patented in 1961 by Mozes J. Lewenstein and others and the patent originally held by *Sankyo*, a Japanese company founded in 1966.

It's tradename is *Narcan* and it was ultimately licensed to *Endo Labs*, today a subsidiary of DuPont Merck.

Naloxone's best feature is, when administered to patients who haven't taken opioids, it has almost no pharmacologic effects.

In patients who've taken opioids, naloxone increases respiration within a minute or two of its injection. It awakens patients who've fallen asleep from opioids, and it raises their blood pressure. It still causes withdrawal in addicted patients, but it saves their lives.

What's more, according to Dr. Avram Goldstein, in his book *Addiction, From Biology to Drug Policy*, "If naloxone is given just before every dose of an opiate, in order to prevent all the opiate effects, neither tolerance nor dependence will develop."

It's a modern lifeguard for patients drowning in opioid drug effects.

There is one downside to naloxone, we can't take it by mouth, only by injection or, more recently, through our noses in a procedure called "nasal insufflation."

The most recent addition to our lifesaving storage shelf is a drug known as naltrexone. It's advantage is that it can be taken by mouth. Its tradename is *Trexan* and it's also available in an extended-release form known as *Vivitrol*.

Like naloxone, naltrexone will prevent the onset of abuse behavior and addiction, but that biological effect runs headlong into human behavior. Continued oral naltrexone will block future opioid use, if, and only if, the patients continue taking it. But often, patients miss the high or the buzz from opiates and search them out. Then they give up the naltrexone and return to their old addictive habit because once they have the opiate, they take it. Human craving is not a function the craved item's absence, but once they have the craved item, if it's available, they're driven to take it.

Like it or not, we're all creatures of habit. It's why addicts in detox or rehab need continued aftercare, either from a psychologist, psychiatrist, group therapy or concerned citizens. Just continued aftercare.

Naltrexone is also being studied in medical laboratories because when physician's lower the opioid-blocking dose twenty- or thirty-fold, it paradoxically helps patients suffering chronic pain due to autoimmune diseases like multiple sclerosis and inflammatory bowel disease. It's used off-label to treat fibromyalgia and diabetic neuropathy.

It does work and several investigators have proven that with blinded studies. In the USA, naltrexone needs regulatory approval from FDA before physicians can prescribe it and its manufacturer needs to manufacture a low-dose formulation.

Buprenorphine is the last drug I'd like to discuss because it too has lifesaving properties. It also returns to the idea represented by *Nalline*, that it's both an agonist and antagonist, however, we've learned how to use it in therapy, and we don't use it to diagnose opioid addiction. Precipitating withdrawal is truly cruel and unusual punishment.

Buprenorphine was developed back in 1966 as a replacement for morphine. It was released to the market by the company then known as *Reckett*. They trade-named it *Subutex* and it was a sublingual formulation, that is held under the patient's tongue like the way heart patients treat their chest pain with nitroglycerine. It's street name was *Bupes*.

Medication-Assisted Therapy (MAT) to treat opioid addiction would take another thirty years.

The National Institute on Drug Abuse (NIDA) asked *Reckett* management to develop a combination tablet. Reckett developed a drug they named *Suboxone*, a combination of buprenorphine and naloxone.

Suboxone has become the chief medication to prevent opioid abuse in patients who've chosen rehabilitation, although to take buprenorphine, patients must not take any other opioid for twenty-four hours so the buprenorphine doesn't produce a withdrawal reaction.

Europe Beckons

I joined *Knoll Pharmaceuticals* as a global project manager in 1998. *Knoll*, pronounced Ka-nole in German, was that pharmaceutical division of the chemistry giant *BASF*. Unrelated to my job change, that same year the death rate from unintentional overdose in the U.S was fewer than 10,000 victims. Six months after I joined them, *Knoll* assigned me development responsibilities for a new formulation of the drug trade named *Dilaudid*. I managed development in the U.S., U.K. and Germany and flew to California monthly too because the company that formulated *Dilaudid* as a once-a-day dose was in Palo Alto. Drug development includes learning how to manufacture the new drug, formulate it, test it in patients to ensure it is safe and effective, register it with regulatory agencies and prepare a marketing plan to develop a market quickly. It's an expensive, multi-disciplinary process that takes a few years and costs tens of millions, whether dollars, Euros or Pounds. I managed that process, hence my travel schedule.

My U.S. division hired a German tutor to meet with me weekly to prepare me for my new duties. There is a side story associated with my learning German. My then wife and I were standing in line at the ticket window at a local movie theater when I heard from behind me, "Herr Gold, wie geht es ihnen?"

She was asking me in German how I was, so I turned, and it was my German tutor waiting in line to see the same movie. I answered reflexively in German, "guten Abend" (Good Evening), introduced my wife just as they opened the theater doors to us, and we all filed in.

By 2015, that unintentional opioid overdose death rate had risen more than four-fold to 44,000 victims. *Dilaudid's* market changed in the twenty-first century too. It has a new life and new names to go with that new life because part of that new life comes from being manufactured illegally in China, not legally in Germany. It became a drug of abuse, traded illicitly, on the street where a *Dilaudid* tablet is known as a Dilly. Others call the drug The Big D, or they refer to the powerful firecracker, and call *Dilaudid* the M-80. It played a big role in those 44,000 deaths due to overdose.

Flashing back to 1998, the noise level in the big jet changed as the air hostess announced, "*Wilkommen in Frankfurt,*" then switched to English and said, "Welcome to Frankfurt." I buttoned the top of my shirt, checked my tie, and watched activity increase in the first-class *Lufthansa* cabin. When I boarded, I handed my green raincoat, that always traveled with me, to that air hostess who hung it in a closet for the duration of the trip.

After we landed, she asked first-class passengers if they had coats and I always said, *Mein grüner Mantel,* which means "My green raincoat." Two minutes later she handed it back to me. Hanging it for me was a nice touch along with free movies, decent food, and good coffee that I sipped while the captain landed the big plane because it was early Monday morning. I was always the first to get off the plane, in a hurry to get to work.

My new job required me to fly into Frankfurt's airport twice a month for the three years I worked for them. I became conversant in German, and I could always manage to get my rental car first by kibitzing in German with the *Hertz* clerk.

As mentioned, *Knoll* was founded in Germany way back in 1886 and almost 90 years later, in 1975, was taken over by *BASF* and in 2002 *BASF* sold the entire division.

I called it quits and retired, in part because my now ex-wife said, "If you want to retire, go ahead, I earn enough for both of us." That conversation is a separate issue.

There were too many Sunday-night flights to Europe that placed me at my German desk at 8:00 AM, Central European Time (CET) or my English desk at 8:00 AM Greenwich Mean Time (GMT). I managed development operations in Nottingham, U.K. also, so on alternate trips I landed in Manchester, UK, worked in Nottingham for a few days before I moved on to Germany. I kept things even that way.

As an aside, I was with Knoll for six months when I suffered a heart attack and then subsequent open-heart surgery. I was only 52 and was back at work three weeks after that surgery. My first day back, I learned my responsibilities were summoning me to Cancun, Mexico to host a group of doctors. I took them snorkeling but the scar on my chest was too raw, so I cut that diving adventure short as I saw a shark eyeballing me. It's no wonder that I thought retirement was a good idea.

Knoll had appointed me head of a project to develop a new 24-hour formulation of *Dilaudid* a name they borrowed originally from the old name laudanum, given to an early opium formulation known as tincture of laudanum, a name borrowed by the English physician Sydenham way back in 1676 when he invented the formulation. Although he borrowed the name from Paracelsus, a Roman physician who lived 100 years before Sydenham, he did not borrow the recipe for the formulation. *Knoll* owned that.

A tincture is any drug dissolved in ethyl alcohol. The name tincture calls up a memory from my childhood. If I cut myself, my mother applied an orange concoction called *tincture of merthiolate*. All I know is that it hurt more than the cut and I've since learned it had mercury in it. Fortunately, my brain escaped injury...from that anyway.

My corporate charge was to develop a slow-release formulation the Board of Directors had named *Dilaudid OROS*, made by incorporating *Dilaudid* into the drug delivery capsule known as *OROS* developed by the *Alza* Corporation. *Knoll* was headquartered in Ludwigshafen, Germany, *Alza* in Palo Alto, CA, clinical trials were managed, in part, by our Nottingham, UK office and my office was in New Jersey.

OROS is a drug delivery technology *Alza* developed, patented, and applied to a host of drugs to broaden their markets. I toured their manufacturing facility in Vacaville, CA, home of a large state prison, and less than an hour-drive from San Francisco Airport. It was where they manufactured the tablets and I learned how they worked. *OROS* stands for *Osmotic Release Oral Delivery System* and *OROS* tablets work by absorbing water from our bodies, but that absorbed water enters only one side of the tablet. The water taken up by the tablet doesn't wet the drug built into the tablet because a membrane divides the tablet internally and separates the water from the drug. The water pushes the membrane, and that increased pressure pushes the drug out through a hole, drilled by a laser, on the other side of the tablet. We've given that mechanism the head-scratching name *zero-order kinetics*. Understanding it isn't as scary as the name sounds. Think of a toll booth letting one car or one truck through each time the driver pays. No matter how many cars are backed up at the toll booth, only one car at a time will fit through the booth and only if the driver pays. That's zero-order.

It's the same with an *OROS* formulation, no matter how much drug the pill holds, or how much water the tablet absorbs, the laser-drilled hole will always let the same amount of drug through the hole – zero order.

The opposite of zero-order is something we've named first-order. Think of my toll booth analogy, if we double the number of toll booths open at a time, the rate at which cars can enter the tunnel will also double – first order.

All that intercontinental running around bestowed on me a million frequent flier points. When I remarried in 2000, we honeymooned in Australia by flying first class to Sydney and staying in a downtown hotel all paid for with my points. Those days are gone. So is the marriage for that matter, but not because of drugs, excessive travel, or frequent flyer points. My marriages are zero-order, but I'll skip completing the analogy. I'm unmarried and single parent our thirteen-year-old four days a week.

Our U.K. colleagues had good senses of humor as well as access to European patent records because twenty years ago, when I worked there, the U.K. hadn't begun talking about leaving the European Union. One day my colleagues must have been curious or simply bored. Someone searched for and found the original 1922 patent for *Dilaudid*, known by the generic name hydromorphone.

During a global team meeting one morning, my U.K. Team Captain pulled a copy of that patent from his briefcase, unfolded it, and said to me, "Look at this," as he handed me the paper. He pointed to a spot on the paper and added, "It's even signed by Kaiser Wilhelm. There's his seal." We passed it around so all the Team members could stare at it, and we put it up on the teleconference TV screen where the Germans could also see it. Then I continued our meeting although I had to expend more effort to keep everyone's attention.

History textbooks teach us that Kaiser Wilhelm abdicated in 1918. Perhaps he hung on for a little while after he abdicated because he liked government, or the patent application was processed after he left. Whatever the reason, the Kaiser's seal and signature are on the *Dilaudid* patent. Maybe it was his last official act. We'll never know. I learned separately and later the German emperor's title *Kaiser*, borrowed directly from the old Latin emperor's title *Caesar*.

But the question is, how did *Dilaudid* change its status as a hundred-year-old imported pain killer to a street drug with a nickname?

It still relieves pain, but in addition to relieving that pain, *Dilaudid* also makes users feel good. That good feeling not only motivates them to take a second dose, only in part because their pain is gone, but that good feeling was so rewarding it became sufficient motivation to take another dose. That good feeling motivates users take a third dose seeking the feeling, then a fourth, and so on. It doesn't take very many doses before users start looking at the clock checking if it is time for their next dose. That's a behavior pattern that marks the beginning of an addiction. One definition of drug addiction is taking a drug to avoid withdrawal reactions and the simplest withdrawal reactions to *Dilaudid* wearing off was the pain returning, but users also missed that good feeling, the feeling of well-being the drug produced. We call that feeling of well-being, euphoria.

The active ingredient in *Dilaudid*, hydromorphone, is an opioid drug like morphine and the chemist who invented it used morphine as his starting material. It works with the same mechanism morphine does, but the *Dilaudid* molecular structure is protected by a patent. Morphine can't be patented because it's a natural product, but synthetic products, even semi-synthetic- can be patented and companies can charge more for patented products. Additionally, they won't have competition for the duration of the patent.

What is that mechanism? Forty-five years ago, scientists worked out how morphine works, and the receptor mechanism is fascinating as I've discussed.

What about that good feeling, the euphoria, *Dilaudid* produces in addition to relieving pain? As with every opiate drug out there, either derived from morphine or made entirely at the lab bench by the hands of a talented chemist, the entire class of drugs produces that feeling we've labeled euphoria. What's euphoria? It's intense excitement, happiness, self-confidence, elation, and relaxation, all without a care in the world. Every good feeling you've ever had, all rolled into a single experience, perhaps just shy of sexual climax. No wonder people begin to look at the clock anticipating their next dose of *Dilaudid* or any other opioid.

There is another phenomenon all opioid drugs exhibit and we call it tolerance. The first hint of tolerance is when users sense that the pain they treat with our *Dilaudid* example seems to return sooner and sooner after their last dose. *Dilaudid* causes tolerance reliably, after as little as a week or two of regular use. Tolerance is well-known in the medical community and, in the U.S. *Dilaudid* is classified by the FDA as a Schedule 2 drug. They define Schedule 2 drugs as having "a high potential for abuse which may lead to severe psychological and physical dependence." So even the law says we should be careful with *Dilaudid*. There are other Schedule 2 drugs, but the best-known member of this class is morphine. By contrast, Schedule 1 drugs have no medical utility and heroin is the best example, a very powerful drug with no medical use.

One of the characteristic side effects of opioid drugs is constipation. Strangely, it's the only side effect to which users do not become tolerant. Go figure.

Does this mean that even if our doctor prescribed *Dilaudid*, we must still be careful? Yes, especially if you're an American reader, because two-thirds of the world's *Dilaudid* is consumed by Americans and that's not because *Dilaudid* treats pain any better than any other pain killer. It's because *Dilaudid* makes users feel that euphoria I described, and that the American medical marketplace is more loosely regulated than the socialized medical markets in Europe.

None of the behaviors I described above is illegal, but they are all, at the same time, behaviors exhibited by addicts who get their drugs from any dealer. Street users treat their tolerance by taking the drug more and more frequently. Prescription users who are already staring at the clock, begin to think it's okay to take the drug a little earlier than the prescribed dose interval of six, eight, or twelve hours. We all say to the mirror, "After all, who will know?"

Sadly, I maintain any time you take a drug more frequently than your doctor ordered, you are abusing that drug. One of the other behaviors exhibited by prescription drug abusers is "doctor shopping." If their own prescribing physician says, "I think you've had enough and your injury is healing nicely," they find a new doctor, whine about their pain, real or imagined, and get a new prescription. That's also drug abuse and the problem with abusing an opioid drug, is the metaphoric grip those drugs seem to have on the user. That's drug addiction and the only control we have is avoidance.

So, the next time you take *Dilaudid*, prescribed by your doctor to treat your broken arm or knee injury, and you find yourself checking how much more time you must wait until your next dose, you'll understand what is going on. Finish your prescription, be uncomfortable for a few hours and get on with your life, drug-free.

It's All in Your Head

Our own opioids, the endorphins were discovered back in the dark ages of 1975 by Sir John Hughes and Hans W. Kosterlitz. In '75, I was still a graduate student in pharmacology. They called their discovery enkephalin, which meant "in your brain." Today we call all the chemicals in our bodies that bind to our opiate receptors, endorphins, a word that can be interpreted as our *brains' inner-morphine.*

I met Sir John Hughes earlier in my career before I went to graduate school and before he discovered "enkephalin," a discovery marked by forethought and hard work that won him a knighthood from the Queen. Way back then, I was a recent bachelor's-level college graduate in the right place at the right time, working as a lab technician for a pharmaceutical company in New Brunswick, NJ. Hughes consulted for my boss, and I met him because my boss assigned me to assist Hughes during one of his visits to our labs. He may not remember, but I do.

What is the endorphins' role in our physiology?

They make us feel good!

Doesn't that make sense? What if they're missing?

As I think about more and more about addiction, I keep revisiting the idea that the drug use that leads to addiction is like a deficiency disease. The receptors that drug binds to lie empty.

We'll look at diabetes because it is the usual deficiency disease that jumps into our heads. Type I diabetics take insulin because their pancreas doesn't make enough of it. That lack of insulin makes their blood sugar go up as a result. Type II diabetics don't take insulin because their bodies may make enough insulin to keep up with their sugar but somehow, they can't or don't use what they make. They take oral drugs that helps their bodies use the insulin they make, or to help them make more.

Either way, diabetes is the classic deficiency disease.

But how does that fit with opioid use?

Perhaps opioid users don't make enough endorphins and as a result, they feel "lousy," or "sick," how most addicts describe their feeling when they don't take their opioid of choice.

Similarly, people use opioids because they like how they feel after using the drugs. They like the "buzz." We have also implicated the endorphins somehow changing how we feel so the same thinking leads me to believe people take opioids to replace their own endorphins because their brains don't make enough of them, or their bodies don't use what they make. Type I Diabetics fight rising blood sugar with insulin and Type II diabetics fight it with oral anti-diabetic agents. Perhaps endorphin-deficient people fight feeling lousy by taking opioids because they're not making enough endorphins or can't use what they make.

Is there any evidence to support my idea?

Yes.

Consider that running increases endorphin levels and endorphins trigger what is called the runner's high. We have all heard people say what amounts to, "When I come in from my run, I feel great, as though I'm floating."

Running raised their endorphin levels. We can think of it as a mild "buzz," without the hallucinations or sleepiness.

We've learned that it isn't only running that raises those endorphin levels, other pleasurable activities increase them as well. We can for example, raise endorphins by eating dark chocolate, exercising regularly, practicing yoga and perhaps the most interesting activity, being with other people. When our endorphin levels go down, we seek activities or conditions that make us feel better by raising those endorphin levels. I encourage readers to seek the company of others.

That brings us to the other deficiency that triggers or maintains drug abuse. The idea that decreased social interaction leads to increased opioid use.

I believe being with other people, socialization, stimulates our endorphin levels because humans are social animals. We like to be with other people, like bees living in hives, humans thrive when they live together. It's the reason cities grew from villages. Aristotle said, "Whoever is delighted in solitude is either a wild beast or a god." Humans are neither beasts nor gods, they're just humans.

We create and live in societies. The problem is Western society is changing. We are decreasing our interaction with other humans, our contact almost driven downward by our choices. It has been going on for a hundred years and my own family is a good example of how even family structure reflects that idea. It is the first, but not the last trigger to decreased interaction.

As far back as I can trace, my father's grandfather- my paternal great-grandfather- had six brothers and sisters. I know that only because I sent my saliva to a DNA testing company and two years later, I received an unexpected email from Belarus, from a family to whom I was related. My point is that my dad's family of origin was huge by modern standards, but 100 years ago when my grandfather fled Belarus, he settled in Elizabeth, NJ. He had three children, half the number his father had. My Dad was one of those and he had an older sister and younger brother. Neither his brother nor sister had children and Dad and Mom had me. So, in two generations, my family went from six brothers and sisters to me.

I tried to reverse that trend by having two kids with my first wife and one more with my second, but I think my own experience represents how family structure has changed even more in the last century. Not only has family size decreased but the number of divorces in the U.S. has risen and has reached nearly fifty percent of all marriages…although 100% of my marriages failed.

The point is, if we feel well, we can thank our endorphins, and if our endorphin levels go down, we don't feel well. It is an intriguing question of why being with other people would raise our endorphin levels. Is the sense of loneliness, an endorphin deficiency?

The recent Covid pandemic worsened our isolation. It made us physically isolate from each other for health reasons. Businesses, schools, and stores closed, we hid our faces with medical masks, and we were told to stay six feet apart. No wonder that fatal drug overdoses continue to increase yearly, especially through the pandemic. We missed out social contact.

Society has changed in other ways during the last century in addition to changing family size. We have moved from crowded urban environments, where we lived in apartments, frequently with more than one immediate family sharing the same apartment, to suburban, single- or two-family homes, each with just one or two adults who raise just one or two children. Families, because of scattering like that, don't supply the social interaction that make us feel good all the time by keeping our endorphins at normal, elevated levels. Similarly, isolated in our single-family homes in the suburban boonies, that new structure doesn't supply the interaction with unrelated people that define living in a society. The net result is that our endorphin levels fall and that makes us unhappy. It's a hypothesis that could be tested.

Correlated with all that decreasing family size and people staring without emotion into their cell phones instead of talking to each other, the rate of opioid usage and death from overdose is still increasing.

There is a danger in that statement because in statistics, it is a mistake to say, "Correlation implies causation." That means simply because two variables change at the same rate, such as family size decreasing and opioid usage increasing, mathematicians teach it's an error to conclude that one caused the other.

Proving correlation takes data and those data exist also although before we can dig into them, we must be aware of the disaster abusing opioids has brought us.

In January 2019, the National Institute on Drug Abuse (NIDA) reported that 130 people died each day from opioid overdose.

I propose we cannot lay blame on anyone or any organization for that death rate even though thinking about why so many people die from opioids stimulates a natural human response for us to want to place blame. Is it pharmaceutical company TV advertising that increased opioid usage or their invention of new formulations? Is it weak regulation of that pharmaceutical industry by the US FDA? Is it people living in smaller family structures or in the suburbs such a cause of unrest that people turn to opioids? Even more cynically, is it people playing games or sending texts on their cell phones that has created unrest and stimulated opioid use?

It is probably all the above. I don't encourage placing blame though, but I do challenge all of us to accept responsibility. Maybe that responsibility is nothing more than sitting with our children for an hour a day and talking to them without interference. Agree in advance, "I won't look at or answer my cell phone for the next hour and neither will you."

It's a start.

Let's look at a little more data. NIDA went on to report that somewhere between 21 and 29 percent of people prescribed opioids for chronic pain, misused them. That shows we don't follow directions very well, even medical instructions. No surprise there, humans are an independent species. As a society, we don't follow dietary guidelines very well nor come to a complete stop at a *STOP* sign if there's no traffic, unless we see a police car waiting to ticket us.

Between 8 and 12 percent of people prescribed opioids for chronic pain developed an opioid use disorder and half of those patients transitioned from prescription opioids to using heroin. That suggests the drugs themselves are a lot more powerful and dangerous than anyone realized a generation ago and their power over us and our danger from them, is only now being understood, reinforced by the statistics.

Let's take a different perspective to emphasize what I am saying. If we reverse the question we ask of those statistics, and ask users, "How did you begin using opioids," we learn that 80 percent of all heroin users began by misusing their prescription opioids. Again, prescription opioids are not the cause of the epidemic, merely an easy, nearly barrier-free route taken by people who expected differently.

The story gets worse if we look at the spread of opioid use. The same NIDA report showed us that opioid overdose in the U.S. Midwestern region increased 70 percent in just fourteen months from July 2016 to September 2017. But before we look at that astounding increase and conclude that opioid abuse is moving to the more rural Midwest, that same NIDA study reports a 54 percent increase in opioid use in sixteen large cities. So, opioid use, and abuse is everywhere and both spreading and growing at a seemingly uncontrollable rate.

I'll say again, they are more powerful and dangerous drugs than we ever realized. That's scary, because as the adage teaches us, "the cat's out of the bag," regarding opioid usage.

Just as the cell phone market exploded because people liked using them, so opioid use exploded because people liked taking them. Cell phone manufacturers saw people buy what they made so they made more to meet the market demand. Opioid manufacturers saw the number of opioid prescriptions in the U.S. nearly triple, from 76 million prescriptions in 1991, to 219 million prescriptions by 2016, just fifteen years, so they manufactured more to meet the market demand.

What drove that increase? Simply economic forces, such as that increased market demand, together with a pharmaceutical industry responsive to that market demand and a U.S. FDA that bought the claim from the industry that their new drugs and formulations they proposed to market would be safer than the old ones. In other words, all the above and not a single factor, but the interaction of all those factors.

Market demand increased because people claimed they were in pain. In fact, there are reports that a condition known as fibromyalgia is associated with chronic muscle pain due to decreased endorphin levels. I'm certain other pains will be ascribed to decreased endorphin levels too. I already proposed that loneliness might be due to those decreased levels. I'll even propose loneliness is a symptom of decreased endorphin levels.

That increased market demand was met by the pharmaceutical industry that made more opioids available. We must realize and accept that our free-market economy drove that expansion in the prescription opioid market. The pharmaceutical industry, like the auto or technology industries, are largely public, meaning many of the companies issue shares that trade publicly in the stock market, and the company's management had a responsibility to increase profits to keep the stock price stable, or try to increase it. That's how a free-market economy operates. The dilemma presented is, how does a company compete in a free-market economy and ensure it doesn't endanger public health at the same time? That's evaded the pharmaceutical industry, in part because many companies followed sales patterns set by other companies. If one or two pharmaceutical industry companies reported increased opioid sales, other companies, either supplied them with raw materials or entered the market with a new formulation of an old opioid to seize the retail momentum begun by the companies forging new sales with their opioid drugs. My leadership of the Team charged with licensing Dilaudid OROS is an example of a new formulation of an old drug.

The pharmaceutical industry had a responsibility to increase market demand by expanding their product lines. They developed new formulations of existing drugs that stretched the time between doses, thereby eliminating doses and increasing convenience. For example, dosing went from two or three times a day, to once a day and the companies claimed that dose schedule was safer. The U.S. FDA accepted that argument, because, after all, the drugs had been on the market for a generation and experience taught everyone, the drugs were safe…if users followed dosing instructions.

But I've also said humans don't like to follow instructions.

So, users took the one-day tablets and crushed them and either snorted or dissolved them in water and injected the contents. They felt great and got a real buzz. That defeated the new formulation, introduced drug abuse behavior, and fed the addiction with a prescription drug.

But if patients finished their prescriptions and found themselves "hooked," they had no choice but to move into the illicit drug market. That's one of the mechanisms by which prescription opioid usage leads to illicit drug usage. We can picture a scenario. Let's say a patient breaks his or her leg. The attending physician issues an opioid prescription to last two weeks. Two weeks later, the patient uses up that prescription and the physician refuses to renew it saying something like, "You're two weeks past surgery and you said your pain is much better."

The only choice for that "hooked" patient is to obtain opioids illicitly. Once again, the drug played a role because opioids are more powerful and more dangerous than we ever thought possible. We are learning, but it took us a few generations and prescription opioids for pain continue to be one of the routes to addiction.

Feels Like a Cold

Your skin seems to crawl, you feel sleepy, your appetite decreases, and you feel weak. Maybe you have a cold or worse, the flu.

Wait a minute.

Energy, appetite and strength are physical expressions that are managed by our brains. If we have a cold, how do our brains know? We've always been taught our brains aren't connected to our immune systems. But if the virus has crept inside us, our brains somehow respond with those symptoms I listed above. There must be a signal, but what is it?

It isn't the virus getting into our brains. Viruses are kept out by something called our blood/brain barriers. We're taught it's our immune system that fights the virus. We make antibodies to viruses, and we have special blood cells whose job is to gobble up those nasty viruses.

But recent research has shown us our brains are connected to our immune systems. So, when our immune systems kick into action, our brains are alerted. It's how we get those creepy, weepy, sleepy side effects. It's why we call the central processing units of our computers, their brains, and why errors that creep in are called computer viruses.

Many of those same symptoms, sleepiness, creepiness, and weakness are also side effects of opioids. That opens another question. Why would people use drugs voluntarily that will make them feel as though they have a cold? They're not pleasurable feelings.

Because the drug's buzz is better than the side effects it produces.

One of the newest discoveries in opioid effects is that our brains are connected to our immune systems. It's work that was triggered by and developed out of the AIDS epidemic in the 1980's. A third of the AIDS patients abused heroin. Scientists began scratching their metaphoric heads and asking, "Is heroin immunosuppressive? Did its use suppress our immune systems and cause users to let the AIDS virus infect them?

Yes, heroin and morphine are immunosuppressive, and our brains do tell our immune systems what to do. The AIDS patients were studied with morphine.

So, it makes sense that a cold or the flu signals our brains to cause side effects. What surprised us all is that opioid drugs, in addiction to activating our brain's opiate receptors, also signal our brain's immune systems.

That opens a host of possibilities. Could our own endorphin/enkephalin systems interact with our brains' immune systems? Do our own opioids tell our bodies how to react to someone sneezing next to us on an airplane or sniffling on us during our morning commute?

Can we treat opioid addiction through our brain's immune system providing another pathway to help addicts?

Scientists around the world are thinking about that. I read a 2014 article co-authored by Dr. Mark R. Hutchinson of Adelaide, Australia and Dr. Linda Watkins of University of Colorado, Boulder. Their article, in the medical journal *Neuropharmacology* was entitled, "Why is Neuroimmunopharmacology crucial for the future of addiction research?" I wrote to each of them asking if there was progress in their research and who else was working on it. They both responded.

Not only do our brains have branches of our immune systems, but parts of our immune system also have opiate receptors. So, our opiate receptors are also found outside our brains.

That opens even more questions. Are opioid drugs immunosuppressive by binding to opiate receptors on cells in our immune system, or do they act on our immune system by binding to our brain receptors?

I'm still confused, and I look forward to scientists figuring that out. The questions are not only fascinating, there is real promise of advancing our knowledge of how to treat opioid overdose.

I dug deeper and continued looking at social behavior because I've stressed loss of social contact played a role in the spread of opioid abuse. It is social contact that passes cold viruses to others too. It's all connected.

I found a 2019 medical paper in the journal *Neuroscience*, written by Kayt Sukel, entitled, "In Sync: How Humans are Hard-Wired for Social Relationships."

She quotes Dr. Simone Shamay-Tsoory from the University of Haifa. "Historically, scientists attempted to understand social behavior by looking at one person at a time. But that is not sufficient to understand the neural underpinnings behind such complex interactions."

What complex interactions? She was talking about empathy. She quoted Dr. Shamay-Tsoori more and the Israeli scientist defined empathy as the "Ability to feel and understand other people's emotions. There are both emotional aspects of empathy, where we share the same feelings as someone else, as well as cognitive aspects, which is our ability to take some else's perspective..." The English expressive is, "Try to consider it from my perspective."

She went on to describe a fascinating experiment in which she and her colleagues did brain scans of two people simultaneously. Dr. Shamay-Tsoori described the experiment. "The researchers recruited people in pairs. One of the two participants received a heat stimulus that resulted in a burn-like sensation. The other held their opposite hand, offering empathy and support. When the pairs were strangers, the researchers did not see much of an effect. But when the pairs were romantic partners, they saw similar patterns of alpha-mu band brain activity, a type of brain wave previously implicated in empathy, in the right frontal lobe in both the person feeling the pain and the person giving the comfort. In fact, the greater the brain synchrony between the two, the less pain the person receiving the heat stimulus reported, suggesting the comforting touch of the empathizer may contribute to a form of touch-related analgesia."

Whew!

Heroin suppresses our immune systems in addition to taking away pain. But we can also provide analgesia to romantic partners simply by touching them. It lends new meaning to the expression, "You feel good to me."

It's a new dimension both in socialization that shows evidence people are wired together, and that our immune systems connect to our brains and open a new pathway for communication and, with luck, therapy.

Is There a Way Out?

We have a public health emergency on our hands and, although I don't mean to paint a grim picture or be a pessimist. We must work our way out of that emergency and it's hard work. We can't, in good conscious, tolerate drugs killing 100,000 young people a year with no end in sight.

Where do we start?

The good news is, I think we have begun. Physicians are less likely to renew opioid prescriptions, and, in many states, it is illegal to renew them. Secondly, the prescriptions they write are for fewer days because most patients don't need an opioid for a full week or more after a minor surgical procedure like setting a broken bone, removing an unsightly wart, or excising a small tumor. Two or three days is sufficient for the pain from those three procedures to let up and many others too. In some cases, opioids may not be the first line of defense against the pain of the medical procedure or temporary relief from a minor injury. Acetaminophen (*Tylenol*), or even simply a bag of ice on the wound, may be sufficient. We're not able to sanction 'laying on of hands' notwithstanding the interesting experiment I described in an earlier chapter. My description does fit with the role of families and social structure.

We are presented with a problem resulting from physicians being less likely to renew opioid prescriptions.

We tend to reverse errors with a full reversal of our actions when less than a full reversal could work.

A pain epidemic among chronic pain sufferers seems to be emerging from reduced prescriptions for opioid drugs. We need to seek a balance there, perhaps guidelines to follow. It's a good project for pain professionals and doctors who treat cancer to address at professional meetings.

Regarding that pain epidemic, a recent article in the medical journal *Pain* concluded, "Discontinuation, whether abrupt...or with dose reduction...significantly increased risk of suicide compared with those with stable or increasing dose."

Suicide is chosen by opioid users with chronic pain prescriptions because whatever it is that causes their pain, is more than they can handle. If you've suffered pain from injury or illness, you can understand.

That suggests, we have to be careful as we approach one aspect of risk reduction in our society, don't reduce prescriptions for patients who suffer chronic pain. They're lifesaving. We need moderation, not reversal.

The pharmaceutical industry, especially after the *Purdue* fiasco with *OxyContin*, will be less likely to develop new opioids or even to reformulate older ones. Scientists agree, we need a new class of drugs that stimulate our opiate receptors without causing us to lose consciousness, giving us a buzz, producing tolerance, physical dependence or, when dosage stops, withdrawal. Maybe we need a drug simply to stimulate our own production of endorphins. It's a tall order but humans are a smart, resourceful species and the pharmaceutical industry once gave us aspirin that now prevents heart attacks, penicillin that treats infections and a whole selection of drugs that prevent seizures, so we know they're capable of brilliant, innovative work. I also think we all agree they should be allowed to make a profit to distribute to their shareholders too because we all enjoy living in a free-market economy. Exporting drugs is a way to bring cash into the country. That's balance of trade.

Similarly, the U.S. Food and Drug Administration is highly unlikely to approve any product extension for any existing opioid or a synthetic new opioid. I think back to my final days in the industry when one of my projects was to manage the development of a *Dilaudid OROS* tablet that would have been prescribed for twenty-four hours. *BASF* sold my division before we applied for a license to market that formulation. We won't see that again although the *Alza* Corporation, the inventors of that *OROS* technology, ran *Dilaudid Oros* clinical trials in 2006 but they were acquired by *Johnson & Johnson* and *Dilaudid OROS* never came to market. The *Alza* division of *J&J* does market a fentanyl patch, known as *Duragesic*, so my fentanyl discussion must include them. We know they made errors because of the huge fines they paid to States.

The influence of cell phones on human interaction will continue to grow as the phones become more versatile and reach more people. My thirteen-year-old son has one and he watches *You Tube* videos unless I intervene. I just find something else for him to do. He has his own workbench in my basement workshop and I'm not only teaching him woodworking, he's my sous-chef in the kitchen also. He makes a mean macaroni and cheese casserole and, as I write, he's in the kitchen making lasagna.

It is interesting that the Internet is paying attention to the idea that we need more contact with other people. We not only have *Facebook*, *Linked In*, *Twitter* and *Instagram*, but dating sites have multiplied and you can even search for a date of any age, sexual orientation, gender or marital status. There are even newer sites with the goal of identifying people with similar interests. *Meet Up* is one I've used solely to find writing workshops although single parenting keeps me from attending those classes in person.

What about the law and opioid use? As I keep repeating, I believe opioid addiction is a medical issue, not a legal problem. Addicts should be patients first, not defendants. Drug dealers are criminals and should be arrested. That position becomes complicated when opioid addicts deal drugs to pay for their habits. We need a public debate and some new ideas on that little piece of sociological overlap.

There is one position that proposes addicts who deal drugs to support their own habit, should be free from prosecution and treated as patients if they enter addiction treatment, stay free of opioids, and cooperate with law enforcement. It's very reasonable and should be our starting point for discussion.

Similarly, technology must help us. Chinese fentanyl continues to sneak into the U.S. through Mexico and it's scary to consider because of fentanyl's potency. The human dose is so small that a shoebox filled with fentanyl is enough drug for dozens if not thousands of people and it's easy to hide a shoebox in a tractor trailer filled with manufactured Chinese products. Heroin continues to be cut with fentanyl and sold as heroin, which changes the heroin from an already dangerous drug, to one that is lethal. Other fentanyl derivatives like carfentanil and alfentanil, are even more potent than fentanyl so the problem is becoming more complicated as more of those derivatives sneak in. Carfentanil is known as *Wildnil* commercially. We need a super-sniffer machine to detect a shoe-box sized package hidden in a tractor-trailer. The challenge is there, and I think we have the technology to meet it. We need an invention whose job it is to sniffs drugs. It's a challenge for the tech industry. I wish I could write code.

Similarly, fentanyl is showing up incorporated into other drugs. In fact, in the U.S. it is found in most deaths due to opioid overdose, if not as the main drug, then as a contaminant of the main drug.

Lastly, I think a mature attitude toward opioid addiction should replace the knee-jerk response many of have when we read, or listen to a broadcast, about the latest deaths due to these drugs. Opioid addicts are not just medical patients, they are victims of the same changes in society that makes us sit in a crowd and stare at our phones. We have drifted away from each other, society has diluted itself, and the result is, many people turned to opioids to fill the void left by lack of social contact. I keep returning to the idea that loneliness decreases our endorphin levels and opioid abuse is our mechanism to occupy our opiate receptors that lie empty. It's an idea that needs testing.

When you feel so lousy that you begin to think of "taking something," you're better off calling a family member or visiting someone. That bad feeling will pass, and if it doesn't or you find you can't take it, call a shrink.

New Voices, Old Problem

When I began writing about opioids, I shared early pieces of my work with members of several non-fiction writer's workshops I had joined, some recently and others, I joined years ago. Several of them said, "Your subject is important, and you handle it well with a good personal background, but you need new voices, especially addicts' voices."

Others said, "Your writing is boring." I withdrew from those workshops.

Adding new voices is a great idea but tough to accomplish, especially the goal of interviewing recovering addicts. I began by writing to *Narcotics Anonymous* and asking to interview some members. They said, "No," and their dismissal included, "Our experience tells us that you might be served better by someone in the Treatment Industry." At least they responded, and what experience did they reference?

Then I made a list of all addiction treatment facilities within forty miles of my Northern-New-Jersey home. I wrote to those that provided an email address, but I had even less response than I had from *Narcotics Anonymous*. Not even one return email from any treatment facility I queried. Not even a referral. Then I called a few others that provided a telephone number. "Someone will get back to you," several contacts said. No one ever did.

I went to my local police department and introduced myself to the uniformed cop at the front desk, "I live in this town, and I am working on a book about opioids. I'd like to meet with a narcotics detective as background for my book." I handed him my business card that just has my name, cell phone number and email address.

He was courteous, nodded and said, "Give me a minute," and he entered a back room with my card.

I was very hopeful but less than five minutes later, he reappeared and said, "Someone will get back to you."

No one ever did.

As big as the epidemic of opioid use and resulting overdose is in the U.S. I've yet to find an addict or a law-enforcement professional willing to be interviewed about it. Stay tuned.

So, I refocused on a different group. A lawyer I knew from the gym to which I'd belonged wrote me and recommended I meet with an addiction counselor he knew. I contacted the fellow and set up a luncheon meeting that began on a friendly note when he said, "Bring your lunch to my apartment."

I did.

His name is Stephen Jay Levy, and he is a psychologist who spent the last fifty years working with recovering addicts and alcoholics and writing books on the subject, one of which I just ordered. He wrote it with Jacqueline Cohen, and they titled it, *The Mentally Ill Chemical Abuser: Whose Client?*

"My opinion about treating addiction, is that users are ill and deserve to be medical patients," I said to introduce my subject to him. I've espoused that point of view before and will continue to stand by it.

Levy agreed with me, and he said, "The only model that works with addicts is the public health model." The public health model views opioid addiction as a treatable illness, not a crime that locks up the addicts as criminals. He has another quote on his business card, "It's about change," and his clinical philosophy stated on his website says in part, "If I treat them as the person, they are capable of becoming, they will become that person."

Addicts are people fully capable of becoming ex-addicts with the right treatment. During our meeting, he recommended three other books to me.

I emailed him a few days later, "Many thanks for meeting with me today. I've already ordered three of the books you recommended. I would still like to add one or more addicts' voices to my book and if you can be helpful in my reaching that goal, I'd be grateful."

His briefest response was, "Go to open NA meetings." NA is *Narcotics Anonymous*. I felt it would be deceitful on my part if I went to one of those meetings solely to interview recovering addicts, but this is the twenty-first century, so I turned toward social media.

I found a group on *Facebook* called the *Addiction Recovery Support Group* and I requested to join.

The group accepted my request to join because many years ago I had joined *Alcoholics Anonymous*. *A.A.* helped me quit drinking and a year after I got sober, I filed for divorce from my first wife. I haven't gone near alcohol in thirty years although I just went through divorce from my second wife, so I don't think my choice of spouses or divorces from them was alcohol related. I don't miss either wife or my daily glass of Scotch either.

I posted a notice within the *Addiction Recovery Support Group* that said, in part, "I would like to hear from you if you are a recovering opioid addict."

A few days later, another member responded to my posting. She said, "Been clean since January 13, 2019. Was addicted to opiates for 6 years." She included some photos of herself, and she was young and attractive.

"Well done. How'd you get past feeling lousy and treating that lousy feeling with opioids?" I responded to her entry.

She answered me an hour later, "I honestly just quit cold turkey and started coming to…a class. My Dr. put me on the right mental health medications, and it helped with the lousy feeling." Her response was my introduction to the frequent presence of mental illness in addicts. It began to change my opinion about drug addiction. I view drug addiction and mental illness on a continuum, although I can't begin to explain which comes first on that line.

"Thank you," I said. "I'm hard at work on a book about opioids and addiction. I never mention names. No fear."

"You can mention my name if you want. I accept what I've done."

She surprised me and I found a new use for social media, meeting and talking to addicts.

A psychologist I know also gave me a name as he said, "He's a social worker and he will meet with you."

A week later, we arranged to meet at a local *Panera*. While I waited at the entrance, I stared at the parking lot and suddenly a car pulled into a handicap-labeled space and the driver climbed out and stripped to his underwear in the parking lot before donning other clothes.

One of the *Panera* salespeople behind me laughed and asked, "Is that the guy you're waiting for?"

"I hope not," I smiled and announced loudly enough that the entire room of bread-eating customers began to laugh.

Then a middle-aged man walked in, wearing a freshly pressed pink dress shirt and matching tie with two id tags hanging from a ribbon around his neck.

"Are you John Corcoran?" I asked.

He nodded 'yes,' we shook hands, stood at the counter, ordered, and then found a table.

After brief introductions he said, "I'm a social worker [at the local hospital] and I work with recovering addicts."

"Are there any addictive drugs that predominate among your patients?" I asked, as I referred to my list of questions. I always do my homework before interviews, and I had jotted down those questions the night before.

"No, they've taken everything," he said without emotion.

"Are there any changes in government or the pharmaceutical industry you'd like to see, or could recommend?" I asked, between bites of my grilled cheese sandwich.

"I'd like to see all of us assume more personal responsibility," he said. "It's not what can they do, it's what can we do."

We ended our little luncheon with my standard offer, "My goal is to interview some recovering addicts. I respect anonymity." We shook hands and took leave.

Two days later, I interviewed psychologist Charles Heller, one of whose jobs it is to work with imprisoned addicts at what used to be known as New Jersey's Rahway State Prison. It's been renamed East Jersey State Prison. He said, "I can't arrange for you to interview recovering addicts because of HIPAA law."

HIPAA stands for *Health Insurance Portability and Accountability Act of 1996*. It attempts to keep our medical records private and although it stands in the way of me interviewing recovering addicts, it's refreshing to see they are being treated as medical patients despite their simultaneous status as convicted criminals. It's a start.

Then a new week dawned, and I arranged to interview a faculty member at the University of Cincinnati, where I'd earned a B.S. so many years before. I set aside three days to drive out there because my memory told me the trip used to be fun. Of course, I was eighteen the first time I drove out there.

It wasn't quite as much fun as it was back then. It was 619 miles to my hotel from my driveway, and the trip took ten and a half hours. When I pulled into the hotel parking lot, my first thought was, *I should have flown. Next time I will.*

I was scheduled to meet Claudia Rebola whom I had contacted a month before after I read an article about her entitled, *Designing a Lifeline* in my alumni publication, *UC Magazine*.

The lifeline she invented is a naloxone dispensing device for which she shared with me her vision, "I'd like to see it available in train stations, the library and any other place addicts frequent."

She keeps busy and was recently promoted to Associate Dean for Research. She has administrative responsibilities now in addition to teaching students, promoting her invention, sitting for TV interviews, meeting with reporters, and responding to my questions. It took me forty minutes to find her office because the campus has grown so much since I was a student. Although the main campus footprint hadn't changed in all those years, it seemed larger than it did fifty years ago, and the hills were steeper too. Funny how geography changes as we age.

The moment I entered her office, I saw an attractive fortyish woman, wearing blue jeans and a leather vest. She had a Spanish accent because she hailed from Argentina and her motorcycle helmet rested prominently on a side table.

"A motorcyclist," I said.

"Yes, I bought it a few weeks ago as a birthday present for myself," she pulled up a photo of her new bike on her computer. Gutsy lady.

I asked her, "What stimulated you to design a naloxone dispenser."

"I design things based on users' needs and according to the question, how can I help them?" She explained. "When I was at *Rhode Island School of Design*, I collaborated with a faculty member at *Brown University*, and we introduced a device we called the *Naloxbox*. There's one in the Providence Public Library."

"Did it use injectable naloxone?" I asked.

"Yes, injectable naloxone, and there is only one dose within each *Naloxbox*."

A quick check of the *Naloxbox* website (www. naloxbox.org) turned up the interesting figure of 56 *Naloxbox* installations are scattered across Rhode Island.

Claudia left "RISDY," as she referred to the school, and moved to Cincinnati's College of Design, Architecture, Art and Planning, or DAAP.

She invited me to view her display in the DAAP lobby and we walked from her office down to it.

"I have to take this display down shortly," she explained as I read the display boards that discussed her new device, she calls it *AntiOD*.

One of the boards reproduced a short poem entitled, *My Addiction*, by an anonymous AntiOD user:

> *I used for happiness and became unhappy.*
> *I used for joy and became miserable.*
> *I used for sociability and became argumentative.*
> *I used for sophistication and became obnoxious.*
> *I used for the friendship and made enemies.*
> *I used for the strength and felt so weak.*
> *I used for relaxation and became anxious.*
> *I used for the courage and became afraid.*
> *I used for the confidence and became doubtful.*
> *I used to make conversation easier only to stutter my words.*
> *I used to feel heavenly and ended up feeling like HELL.*

More than any other words about addiction, that poem captures the fundamental problem with opioid use. The drugs consistently deliver an effect opposite to what users expect the drugs to deliver, but by the time users recognize they were stiffed, they're addicted to the drug.

"You're assuming personal responsibility," I told her. "It's what John Corcoran said to me last week."

She nodded and went on, "I'm having trouble sourcing naloxone for my AntiOD devices. The Ohio Board of

Pharmacy says I need a doctor's prescription to dispense the drug."

"You've switched to a nasal delivery?" It occurred to me that it's yet another case of laws needing to change, especially if change is to help people.

"Yes."

"And your goal is to save lives," I muttered.

She nodded 'yes,' and added, "Ohio has the second highest death rate from opioid overdose."

That afternoon I left for Pittsburgh where I had morning meeting with John Temple, author of *American Pain*. He is a tall, thin fellow and he met me at the door the next morning. He had red hair and matching red beard.

Our entire meeting lasted 30 minutes but he did theorize why the West-Virginia-Ohio-Pennsylvania tristate region had the highest rate of deaths due from overdose in the U.S.

According to the U.S. Center for Disease Control (CDC) website, deaths in West Virginia in 2017 due to drug overdose was three-times higher than deaths due to firearms. That ratio held for Ohio and Pennsylvania too. Deaths from overdose are much bigger public health and safety problems than firearms in the U.S. Politically, Many Americans focus their public policy arguments on gun control and that puts them at odds with those who defend the Constitution's Second Amendment. That argument is beyond the scope of this book although the data continue to suggest they aim elsewhere...that is, focus their policy arguments on opioid abuse and the public health issue created by the drugs.

The CDC website goes on to cite a national statistic for 2017, 70,287 deaths due to overdose or 19.7 deaths per 100,000 population, up from 13.1 deaths per 100,000 population in 2016, just one year before. Americans kill themselves with drug overdoses more than they shoot each other or drive into each other. In fact, overdose deaths now exceed the combined total of deaths due to guns and accidents, notwithstanding the recent news reports of mass shootings.

By 2020, we witnessed Coronavirus killing 180,000 Americans as I write and, at first glance, that's almost double the deaths from overdose. It's easy to overlook that deaths from overdose not only repeat every year, but they also increase every year. Deaths from Coronavirus has a limited lifespan, so to speak.

I sat with author John Temple during our short interview (presented in a later chapter) …it ended when he began looking at the time on his cellphone…I asked him, "Do you think the mining industry has anything to do with the death rate from overdose in West Virginia, Ohio, and Pennsylvania."

His response surprised me. "That tristate region always had a labor force that worked with their hands. If they didn't work, they didn't get paid, so they couldn't let minor injuries interfere with their ability to earn a living. If they got hurt, they went to their local doctor who prescribed an opiate so they could go back to work without pain."

He had my attention and I observed, "Aren't most of those jobs in China now?"

He nodded agreement and went on, "The population was used to getting an opiate prescription whenever they hurt themselves and that attitude didn't change as business moved abroad. It's why *Purdue Pharma's* sales staff targeted this tristate area."

That's the company that marketed *OxyContin* and triggered the opioid epidemic.

I closed our meeting with, "I'd like to count you as part of my network as I continue to write."

He nodded agreement, we shook hands and I drove the next six and a half hours to my house.

The other not-so-obvious fact in comparing death from coronavirus compared to death from opioid overdose is that coronavirus is more lethal among older people – my generation included- and opioid overdose is more widespread among younger people.

That's more than 100 people a day dying from drug overdose and the economic cost to the U.S., including healthcare costs, criminal justice and lost productivity is in the billions of dollars. To say death from drug overdose is a big problem is the same understatement as saying a coronavirus infection is like the flu.

Oklahoma is OK

Fifty years ago, *J&J* ran an ad for *Band-Aids* that labeled them, "The Generous Bandage." We all know *J&J* is *Johnson & Johnson*, they still make *Band Aids* and they are still headquartered in New Jersey.

A recent *CNN* headline announced, *Oklahoma wins case against drugmaker in historic opioid trial.* Drugmaker? *J&J* has transformed itself, in the last fifty years, from a *Band-Aids* manufacturer to a pharmaceutical company that also still makes *Band-Aids.*

Oklahoma won $572 million from them. More than half a billion dollars.

How did the company that makes "The Generous Bandage" run afoul of Oklahoma like that? That *J&J* verdict comes on the heels of the *Purdue Pharma* story that continues to dominate the news and has for almost a decade. *Purdue* is accused of marketing *OxyContin* too vigorously in local economies that were vulnerable, like the U.S. mining industry in the tristate area of West Virginia, Ohio, and Kentucky, thereby causing the opioid epidemic. I also read a recent *NY Times* piece entitled, *Major Drug Maker is Close to Settling Case to Avert First Federal Trial in Opioid Crisis.* They discussed *Mallinckrodt* Pharmaceuticals agreement "to pay $24 million to two Ohio counties," and "also donate $6 million worth of drugs…"

If winning money from pharmaceutical companies doesn't decrease overuse of opioids, I read a news article in the NY Times, "A federal judge on Wednesday [August 2022] ordered three of the nation's largest pharmacy chains – CVS, Walgreens and Walmart – to pay $650.5 million to two Ohio counties, ruling that the companies must be held accountable for their part in fueling the opioid epidemic." The article went on to say, "The ruling is the first by a federal judge that assigns a firm money figure against the pharmacy chains for their roles in the opioid crisis."

We found is easy to win money from pharmaceutical companies and drug stores. It caused some people to think they'd punished those responsible. I've said before, the pharmaceutical industry didn't cause the opioid epidemic. They didn't help, but they didn't cause it. A weakly-led FDA didn't cause it either. Neither did a thinning of our social structure. They all contributed but no one industry or organization caused it.

We struggle mightily with the spread of opioids, their abuse and deaths from overdose and we have a natural tendency to blame someone or something for that spread, collectively to point our fingers elsewhere. The companies that manufacture, distribute, and sell opioid drugs have money and they've become targets, but I stress again, it's not their fault alone. I maintain weak government regulation and poor prescribing practices among health professionals are all to blame. Blame doesn't solve any problems; it just assigns monetary damages. The reasons for the opioid epidemic are manifold and, as I've said, include poor prescribing practices, too many opioids on the market, a thinning of society, smaller families, and loneliness. It's time we looked in the mirror and, instead of blaming anyone, take the lead. Call your congressional representative, write letters to the editor of print publications, start a blog and use social networking. Speak up!

Oklahoma went after *J&J*, because *J&J* owns a division known as *Janssen Pharmaceutica* that sells a prescription patch of the opioid fentanyl that carries the tradename *Duragesic*. Furthermore, *J&J* is accused of manufacturing and selling many of the intermediary chemicals that other companies used to manufacture opioids. Obviously, companies with that product line must be at fault for the spread of opioid abuse, aren't they?

No, they aren't! They are not solely at fault so we can't point our finger at *J&J* and say, "They shouldn't have sold *Duragesic*, or supplied chemicals to other companies," because blaming *J&J* like that is analogous to my second ex-wife blaming me for her suffering depression that led to her suicide attempt, the breakdown of our marriage that followed and my assuming custody of our son. She always said, "You didn't make me happy enough, so I tried to kill myself."

Both accusations are a stretch, an attempt to blame someone else for our own problem. Divorced men and women are nodding agreement right now. We've all been blamed for our spouses' issues.

Let's think about what I call vulnerable economies. For example, the states I introduced above, West Virginia, Ohio, and Kentucky, fall first, second and third when we rank them according to their rates of death from opioid overdose. They all touch each other, so they are obviously in the same geographic region, and they fit within the larger region we've always referred to as Appalachia because of the mountain range that runs through them or nearby. Mining and agriculture dominated their economies for centuries.

Maybe we are tempted to think, "But they correlate, don't they, states with the mining industry abuse opioids more than other states?"

There is a concept in statistics that applies, "Correlation does not mean causation," or in plain English, just because two variables, like mining and opioid abuse, change at the same rate, doesn't mean one caused the other.

I can give you an example of "Correlation does not mean causation," with another statistic.

Ice cream sales and murders in the U.S. is another example of correlation, not causation, because ice cream sales increase as the murder rate goes up. We are tempted to say, "they are correlated." It would also be easy to conclude, "Ice cream causes people to commit murder," but both would be wrong because of that math principle I mentioned above." Blaming a rise in ice cream sales for the increased murder rate is wrong because neither caused the other.

The culprit is warm weather which drives up both ice cream sales and the murder rate too. Ice cream sales and the murder rate both rise and fall because of that weather, but chocolate ice cream cones didn't drive up the murder rate, hot weather caused both to increase.

I arranged to interview John Temple, author of the best-seller, *American Pain: How a Young Felon and His Ring of Doctors Unleashed America's Deadliest Drug Epidemic.* He is a professor at West Virginia University's Reed College of Media, although he asked me to meet in Pittsburgh where he was vacationing.

I asked him "Why do you think West Virginia, Ohio and Pennsylvania are ranked one, two and three for deaths due to overdose?"

He explained, "For decades, workers in those states worked with their hands and they couldn't afford to miss work due to injury. So, if they were injured, local physicians always prescribed opioids so the pain of their injury wouldn't interfere with their ability to earn a living. That's why the prescription opioid market in those states built a pattern of opioid drug use that led to the highest death rates from overdose." The population had, for generations, been prescribed opioids so they could continue to work.

What he said sounded plausible although I don't know if his explanation has been tested so we must accept it as possible, but not proven.

I believe *J&J* sales of *Duragesic* and drug manufacturing intermediary chemicals did not cause the opioid epidemic, any more than *Purdue Pharma*'s or *Mallinckrodt*'s aggressive marketing did. That's not the same as saying I encourage their corporate behaviors or that I'm in favor of those behaviors, only that their corporate behaviors didn't cause the opioid epidemic. They didn't help either, but we still can't blame them because society is more complex.

We shouldn't lose sight of the important role corporations play in our free economy. The pharmaceutical industry provides 800,000 jobs in the U.S. They also create investment opportunities that help drive the economy. *J&J* and *Mallinckrodt* trade on the New York Stock Exchange demonstrating how our free-market economy operates and the role of corporations in that free-market. We want them to be profitable because of those factors. I mentioned, they sell drugs abroad and that brings money in, helping our balance of trade.

So, if their corporate behavior didn't cause the opioid epidemic, what did? Was it the FDA?

The FDA website has the following quotes, in addition to others, on their first page:

> *The Food and Drug Administration is responsible for protecting the public health by ensuring the safety, efficacy, and security of human and veterinary drugs, biological products, and medical devices;*

> *FDA is responsible for advancing the public health by helping to speed innovations that make medical products more effective, safer, and more affordable and by helping the public get the accurate, science-based information they need to use medical products and foods to maintain and improve their health.*

Protecting public health is a big one and when we look at the death rate from opioid overdose, we must conclude FDA failed, in the case of opioids, to protect public health. A decade ago, there was statistical evidence that the overdose rate was rising due to opioid use in the U.S.A. FDA never stepped in to ask what was going on. They didn't cause the increase in the overdose rate; they just weren't proactive, or even reactive. I believe it is appropriate, maybe even long overdue, for U.S. regulatory agencies to be more assertive. It's why those agencies exist, supported by our taxes. If they say their purpose is to "protect public health," their unwillingness to step in as the evidence mounted, is a failure on their part. Again, not a cause of the opioid epidemic, but a contribution to it because they didn't try to stop it. They can date Oreo cookies, they can follow a statistical trend too. Deaths from opioid overdose have been rising for a decade or more.

For example, *National Public Radio, NPR,* reported in May 2019 that FDA was blocking their approval of a drug known as *Brixadi,* which acts similarly to methadone in that it blocks the action of opioids if the patient taking the *Brixadi* were to try to use an opioid while taking the blocking agent. *Brixadi* is an extended-release formulation of buprenorphine, and it has finally received tentative FDA approval.

I'll cover the issue more in my Lifeguard chapter.

My point is, I believe the U.S. has suffered an epidemic of opioid use in the last decade because of all those factors and more. The combined effects of aggressive corporate marketing, a lack of strong leadership by FDA, a quirky population that doesn't listen to reason, a lack of moral leadership by physicians and demographic shifts. I also think drug advertising on TV played a role and an ineffective Congress all worked, or more precisely, didn't work together.

That's not the same as saying, "It's no one's fault,"

I'm saying, "It is everyone's fault."

You may be quick to react and say, "I don't use drugs, how can I help?"

Start by writing to your Congressional representative and say you want stronger laws, no more drug advertising on TV and no additional opioid drugs approved unless they provide a proven advantage to the ones already on the market today.

And that's just the beginning because writing to your Congressional rep is not only being involved, but also telling the rep that you want action. It is a giant step toward answering the question, "How can I help?"

Be active, not passive.

Marketing, Dealing Drugs

Americans and New Zealanders speak English because both countries were colonized by Britain. In fact, the U.S. was colonized by Europeans more than a century before New Zealand. Both countries have something else in common too, and it relates to the drug industry. They are the only two countries in the world that permit prescription drug advertising on TV or radio. It is called direct-to-consumer, or DTC advertising.

I think both countries should cease DTC advertising of prescription drugs on TV, and I'll tell you why I believe that.

Reporters have asked drug company representatives, "Who do you target with those TV ads?"

Their responses were reflexive, "Physicians."

Yet operationally the opposite is true, DTC advertising places consumers in charge. It arms them to ask their physicians, "I saw [*name-a-drug*] advertised on TV last night. Can you write me a prescription so I can try it?"

We've forgotten that almost thirty years ago, in 1995, even the *American Medical Association* (AMA) recommended banning DTC prescription drug advertising although their reasoning surprised me. They concluded that costs for new drugs were driven up by the cost of the ads. Maybe they were partly correct, one of my prescriptions costs me $100 a month after insurance.

We have the dubious benefit of seeing those ads on TV any time we watch a show and seeing the ads drives up consumer demand for new drugs…even when the drugs demanded are not appropriate for what ailed the avid consumer and the drug advertised was not pronounceable.

Prescription drugs should not be subject to consumer demand. They are to treat illness, pain or recovery, not to satisfy a consumer craving.

The AMA left unsaid, the more glaring conclusion, that they were also trying to prevent physicians' obvious loss of control by putting consumers in charge of their new drug choices. The *AMA* tried to keep physicians in charge of deciding what drugs their patients' need. After all, physicians are trained for it and consumers aren't. It's even worse now because many of the drugs advertised on TV are products of genetic engineering. Many of those drugs have at the end their name, the letters *–mab-* that stand for *monoclonal antibody* and very few of us can judge the appropriateness of products that high-tech and recent. I can't even understand them with all my training. I *Google* them too.

In fact, the more I think about or see TV advertisements for drugs I've never heard of, the more I am surprised to learn how recently TV drug advertising appeared. It was 1997, two-years after the *AMA* said they didn't like it and recommended against it.

Furthermore, the original idea to advertise drugs on TV was tried six years before the AMA's statement. Back in 1981, the British company *Boots* Pharmaceuticals ran an ad on TV in the U.S. for the analgesic ibuprofen, known by the *Boots* tradename as *Rufen*. Within two days, the U.S. government told the company to take the ad off the air and it took sixteen years for drug ads to reappear on TV. *Boots* had gumption though, they bought their advertising time without asking any government or regulatory agency for permission. Those days are gone, although I've been accused of behaving that way too, "Ask forgiveness, not permission," seems to be my motto. Must be why I'm single.

Although it is important that I allow myself some subjectivity and view DTC drug advertising as violating my sense of what belongs on TV, it is equally important that I think objectively about the industry that employed me for so many years. As I keep repeating, the pharmaceutical industry contributes to the infrastructure of our capitalist society, as well as to that society's health and well-being. It was certainly my source of employment and my investment vehicle as much as I helped create pharmaceutical products that are useful. In the U.S. alone, the pharmaceutical industry employs more than 800,000 people and I mentioned my own experience is a good example of their role as an investment vehicle. Years ago, I was awarded stock options by my employer, a Philadelphia-based, former "big-pharma" company since merged out of existence. I exercised those stock options, sold the underlying shares, and used the money to pay off debt remaining from my first divorce, buy an old farmhouse on the Philadelphia Mainline and move three generations of my family, including my folks, my older son and me, into the home I created. Clearly, I value family although I don't necessarily define it classically.

Many of us depend on those pharmaceutical industry investments and we want those investments to pay off. Accordingly, when good news is reported about the companies it drives up their stock prices making our investments more valuable. Increasing sales is good news.

Paradoxically, that is also why the idea drives drug advertising on U.S. TV, because it is an important mechanism by which pharmaceutical companies ensure their new drugs are profitable. But profitability in the U.S. market is also competitive because it is the last big market in the world in which drug prices are not set by a governmental agency. Conversely, the European medical market is entirely socialized, that is most medical costs are paid by government agencies, and within that structure, drug costs are set through negotiation with drug companies before the governments license the products for the marketplace. In the U.S., drug prices are set by the company. There are recent changes to the law regarding pricing but the focus of the law is on very few drugs.

Drug costs in the U.S are largely set by the pharmaceutical companies because delivery of medical care in the U.S. is still largely private enterprise, with a few exceptions. For example, veterans, retired from their service, have their prescription costs paid by their Veteran's Administration pension plans just as they did when on active duty. Medicaid is also a mechanism to bring costs down for a limited population. The rest of us pay full price.

What's my point? Am I advocating socialized medicine? Perhaps eventually we'll move in that direction, but in the short term, it's an unrealistic goal, although physician groups are consolidating so the industry is conglomerating on its own for business reasons. I'd like to see the FDA raise its scepter a bit and regulate opioids differently. But that also suggests Congress must be involved in regulating the drug marketplace as well as in pricing and reimbursement.

The entire structure of paying for medical care, DTC advertising, FDA regulation and financing needs renovation in the U.S.

A *NY Times* article by Neil Vigdor entitled, *It Paid Doctors Kickbacks. Now, Novartis Will Pay a $678 Million Settlement.* He reported that the Swiss company *Novartis* was fined because they used, "...an expensive kickback program for nearly a decade..." That's a lot of Swiss cheese.

What's a kickback program?

Wikipedia defines it as, "...a form of negotiated bribery in which a commission is paid to the bribe-taker in exchange for services rendered."

Negotiated bribery? We all know bribery is illegal and Novartis paid a $678 million fine for their criminal behavior. But Novartis didn't sell any opioids in their scheme. *Purdue Pharma* and *Insys* did.

Recent reporting about Connecticut company *Purdue Pharma* indicts their marketing as too aggressive. They concentrated on rewarding physicians who prescribed *Purdue*'s *OxyContin*. But let's not forget that *Purdue* was a profit-making capitalist company in a free-enterprise nation that provided jobs in addition to a healthy return on investment (ROI).

So, some of us can look at *Purdue*'s track record and say, "They marketed *OxyContin* too aggressively." Others can look at their track record and say, "They succeeded in a competitive marketplace."

I look at their record and say, "Where was our FDA?" Maybe I should extend that and say, "Where was our government?" In a *N.Y Times* article written by Austin

Frakt (April 13, 2020) he reported that *Purdue* focused their marketing on states with lighter regulation of prescriptions. We can look at that tactic and all say, "That's deceptive marketing." They tried to do an end-run around regulation.

That deception wasn't limited to *Purdue Pharma*. Four years before I expanded this chapter, a Massachusetts Federal Prosecutor reported that executives at *Insys Therapeutics*, "led a nationwide conspiracy to bribe medical practitioners to unnecessarily prescribe a fentanyl-based pain medication and defraud healthcare insurers…" Six *Insys* executives were indicted on Federal charges.

I challenge the pharmaceutical industry to state their morality and live up to their statement. I challenge shareholders and private equity firms to demand that. Free exchange of ideas is vital.

Classically, pharmaceutical sales were led by salespeople known as "Detailmen," notwithstanding the obvious sexism built into the title. A Detailman carried his big sample case filled with new drug samples so he could distribute them to his physician customers. Many of us who are old enough can recall our physician opening a drawer and saying, "Try this new drug. Here's a sample I got, and I'll give you a prescription before you leave today."

It's easy to see how *Purdue*'s tactics derived from those earlier sample cases. Where do we draw the line? There is a world of difference between a sample of a new antihistamine or antibiotic and touting a new Schedule 2 opioid. That's why we've all indicted *Purdue*, but I also still fault our government for its evident passivity.

Where do we go from here?

We have too many Schedule 2 opioids on the market to satisfy our medical needs. Marketing drove the increase in the number of opioids on the market to exceed reasonable variety and competition. Companies developed new opioids because they could be patented and patented medications can be priced above generic drugs, that is those that have no patent protection.

Secondly, we would benefit from analgesics that don't show tolerance or produce physical dependence. That's going to take some scientific invention.

Scientific innovation that is on the horizon

What Can I Do?

That's such a common plaint and yet the idea leaves most of us scratching our heads with a sigh, doing nothing, and feeling powerless.

Our tone says, "Oh, enough already," after reading the paper or listening to the TV news in the background every day. We see or hear the statistics, but, for two years, Coronavirus news drowned out every other topic, although, as I write, it was replaced by "Russia is devastating the Ukraine." Perhaps we even know someone who is an opioid addict, or worse, we are a parent in agony because his or her child is an addict. No matter how much or how little we are in contact with the addiction epidemic, we are left with the thought that we can't help, or the self-centered feeling of, "Glad that's not me," as you shut off the TV, put on your uncomfortable mask and head out to run errands.

We've established that addiction is a disease, granted, a disease caused by a drug, but still, like all diseases, one in need of medical treatment. Addicts are human beings who need our help so let's consider what each of us can do.

Sometimes you must stretch to help. I stretched recently. I picked up my ex-wife and drove her to her home when she was released from a hospital. She thanked me profusely, but I responded, "I didn't do it for me or you, I did it so Richard (our now-13-year-old) could be with you for a few hours." He had missed her during her hospitalization, which had nothing to do with drug addiction.

Reach outside your social circle, learn about others and don't be afraid to step up to help.

For example, if you have a relative or a child addicted to an opioid, your first job is to communicate, "Let me help," the same way you would if someone close to you was sneezing from a cold and ran out of tissues. You'd bring tissues to the cold victim and say, "Here, For your runny nose." Similarly, if the person you lived with wore a leg cast and couldn't get off the couch without his or her crutch that had fallen just out of reach and you say, "I'll get it," as you walked over, bent to pick it up and handed it to your couch-bound housemate.

How do you communicate like that with an addict?

Start by avoiding judgement and doing a little homework. Read and learn about addiction, especially in your region, and then find all the addiction treatment centers nearby by Googling them. Then, continuing to ban all judgmental opinion from your mind. You can't be helpful and be judgmental at the same time. What I'm suggesting is an active participation on your part, not a passivity that includes, "You got yourself into trouble, now get yourself out."

If I can pick up my ex-wife and drive her home from the hospital, you can help an addict, especially one in your family, find a rehabilitation facility.

Then begin your conversation with, "I'd like to help you. I've identified a few local addiction treatment centers."

"I just feel lousy," or some similar sentiment is the answer you're likely to hear because many addicts say they take drugs to take away their bad feeling. The problem they have is identifying what is bothering them and talking about it.

"Let me help you get some help," should be your answer. Be gentle. It's important that addicts motivate themselves to seek help. It's equally helpful if you treat them with the respect you'd offer any person.

Remember, you're exhibiting leadership but swimming upstream. Society is fragmented, family life is thin, our homes are likely to be suburban without a village environment, where we watch TV alone and then drive to work alone in our cars or sit on a bus staring at our cell phones and avoiding the rider next to us who may be sneezing. Our daily human interaction is likely to be with that TV or a computer screen rather than with a person or a group. Even in public, many of us don a pair of *AirBuds* or similar gadget and listen to something on our mobile phones as we ride a bus or a train to work.

I'm not immune from that fractionation. When I started this piece, I was at my dining room table sitting across from my son Richard. He was in a *Zoom* class with his school because his school wasn't meeting, closed by coronavirus. He spent time everyday complaining about why *Zoom* wouldn't boot up on his laptop.

Are my son and I working together? Only in a modern sense that we're both sitting at the same table staring at separate computer screens. Sure, I can help him focus when he's distracted by something more interesting outside when he starts looking out the window. He's also wearing earbuds, as he listens to an online teacher, so we don't really communicate. We seem engaged in the same task but not really engaging in a family function. We're modern in another way also…I'm a single, custodial parent most of the week.

It's easy to blame societal fragmentation or small family size for our modern epidemic of drug addiction, but those factors aren't sufficient either. Small families are also a factor but not causative in drug addiction. For example, if there is no family member nearby, then there is no one to whom to turn when that "bad feeling" sets in. Maybe that bad feeling is nothing more than loneliness and drugs treat it by removing all feeling. It's a sledgehammer approach but it works, like killing your entire lawn to rid it of dandelions. All you really need is a hug.

I repeat the idea that extending your desire to help might include reaching out to your Congressional representative. That's active and may be rewarding.

How do you find your Congressional Representative? Go to the following website:

https://www.house.gov/representatives/find-your-representative

and enter your zip-code. The site will deliver to you your representative's name. Then you can begin to communicate with someone elected to bear responsibility.

Perhaps begin with the following question: "Are there Federally-funded addiction treatment sites within our district?"

If the answer comes back, "Yes," then your follow-up should be something like, "Can you help me place someone in there?"

If you're so inclined: "Can you put me in touch with someone in your office who is active in addiction treatment so I can become involved?"

We all know people who "enable" behaviors. I'll take a slight tangent to demonstrate what that behavior is. Richard makes smoothies for breakfast, and he puts his yoghurt-covered spoon on the countertop. I say every day, "Please rinse your spoon so you don't dirty the counter." He learned not to rinse his spoon because his mom would wait until he finished his smoothie and then clean up after him. She enabled his sloppiness which is the wrong behavior. I also find tools on my workshop floor and un-vacuumed sawdust down there so it's a widespread issue in my house. It's also normal for an early teen to be sloppy.

Hence my suggestion. Ask an addict you'd like to help, "How can I help you, help yourself."

I mentioned social interaction and families. We used to have large families and those family members were available to help each other. When I was eight years old, decades ago, I lived in a house my grandfather built for my mom and her family. My uncle, Mom's brother and his wife and their daughter, my cousin Ellen, lived next door in house he built for them. Our grandparents lived three houses away. It wasn't uncommon to hear my mother say, "I'm going shopping, please walk down to Nana and Pa's for lunch?" I would walk through three backyards to their home.

In contrast, today I live alone with my son and the only other relatives I have are my two older children and my cousin who still lives, 200 miles from me, in the same house built by our grandfather. Our society and our families have fragmented, and the result is our families are smaller, far more mobile and society is thinner. There's no longer a tendency for families to live near each other. Once again, I'm not blaming our mobility and small family size for our drug addiction epidemic. I am saying it's a contributing factor but not the sole cause.

Back to your addicted relative. Find an addiction treatment facility and visit it. Is there a fee? Is it covered by your insurance? Is there a waiting list? Those are just a few of the questions you might ask. Some others could begin with, "How long is your recommended course of treatment?"

When you find a place that seems to meet your relative's needs, another conversation with him or her could begin with, "I found a place that could help you and I visited it. I'm eager to take you there and show it to you when you're ready."

"I'll think about it," your addicted relative might say.

"My offer stands. When you're ready, I'll take you there."

That's not only active participation, but it's also offering to help someone about whom you care. Renew your offer often but don't nag, it's not punishment or banishment, it's actively helping.

So, your passive plaint, "What can I do?" has become an active helpfulness.

Our Future is Biased

Pharmacologists are taught a principal, almost a tenet, that morphine and the opioids, in fact most drugs, are carried everywhere by our circulation and anywhere a drug finds receptors it fits, it binds to those receptors. As I explained, some of the cells to which morphine and the opioids bind, stop transmitting a signal that says, "That hurts" to our brains That's why we use opioid drugs, because they stop pain.

But morphine and the opioids have effects that we don't want such as tolerance, that is the need to take a bigger dose to get the same effect the last dose gave us. They produce physical dependence, that is the need to continue to take the drug to avoid withdrawal reactions. There are other effects we don't want either such as constipation, respiratory depression, and falling asleep.

What if we could develop drugs that would only tell the cells that transmit pain to stop transmitting that signal? That new class of drugs wouldn't communicate to the cells that tell our bowels to slow down, slows our respiration, or tells us to take more opioids. The unwanted effects of opioids would be missing.

That new class of drugs may be on the horizon.

I was on *Facebook* a few months ago and I came across a familiar name, James W. Barrett. I had a colleague with that name, and I recalled he was a psychologist. The *Facebook* listing said the fellow who called himself James Barrett was working on drugs of abuse in Philadelphia and given the fellow I knew had a background in psychology, it seemed like the same guy, so I turned my research skills to use, found his email address and asked him to lunch.

Two weeks later we met at a place he recommended near his Philadelphia home. I was an hour early because I had allowed for traffic that never appeared, so I walked around the block. Fifteen minutes before our noon appointment, I decided to get our table, review my questions, and think of new ones. I entered the restaurant and said to the receptionist, "My name is Barry Gold, and I am scheduled to meet Jim Barrett for lunch…"

Before I could finish my sentence, she said, "I think he's here…that's him over by the window." She pointed in his direction where he had overheard us and began to stand.

I walked over and we greeted each other. He hadn't changed other than the effects fifty years has on a man's face. I didn't recall that he was a couple of inches taller than me although his grey hair and my grey hair now match. We both keep active.

We behaved like gentlemen, shook hands, and sat facing each other. He volunteered, "I was slow getting back to you when you emailed me because I was in the hospital for a week. I use a cane now," as he pointed at it lying on the seat next to him.

I initiated my side of the conversation with, "Jim, remind me where our careers overlapped," as I smiled and looked him over as he did the same to me. I regretted complaining about his slow response to my email-out-of-nowhere because I didn't know he had been hospitalized.

He smiled and said, "we were at USUHS together and then Wyeth." He established that we had worked together at two different places. All I recalled was that we'd worked together earlier in our respective careers. USUHS (*Uniformed Services University of the Health Sciences*) is the U.S. Defense Department's medical school, in Bethesda, MD, for training physicians to serve in the military and it was my first job after my postdoctoral studies at *Yale*. Our conversation reminded me how I commuted to work in Bethesda by bicycle from Kensington, the adjacent town. Then I stored my bicycle, leaning it against an oxygen tank in my lab before I showered in the locker room and ate breakfast in the University cafeteria. Strange what memories we conjure when reminded. Maybe, like the rest of the Boomer generation, I just miss my youth.

It was a long time ago. My second child was born the year before I went to USUHS and now he's 46 years old.

We laughed about the answers to a few more questions, such as, "how many kids do you have?" Then he asked, "Have you heard of biased agonists? I consult to a company called *Mebias* and their plan is to develop a biased agonist, hence their name, pronounced Mee Bias."

So, the concept he presented, biased agonists, brings my opioid saga from an explanation of past use and abuse to a hint of the future where abuse may be, forgive me, a thing of the past. I concluded, only partly tongue-in-cheek, that pharmacologists are the only scientists who claim introducing bias can make the products they study safer and more exciting. Did I just invent a concept I could call biased objectivity?

But what is a biased agonist?

I began reading scientific literature about them because the concept is complex and I had to study them before I understood their promise. As I studied them, I appreciated how elegant the approach is.

Treating severe pain, such as from a broken limb or even worse, cancer pain, is a humane response to someone's suffering. The problem until now is that opiates and opioids, while very effective analgesics, have been so easily abused by the public, and that abuse has spread dangerously wide and become a medical and legal problem. I am reminded that two years before I began this chapter, 70,000 Americans died from opioid overdose, and now two years later, it has continued to climb into the hundreds of thousands.

That makes the Holy Grail of our search for drugs to treat that severe, debilitating pain is to find drugs without any abuse liability. That means drugs that don't produce the good feeling, the buzz that opioids elicit; the feeling that makes us mutter, "I wonder if it's time for my next dose?" Or grumble, "Not for an hour and a half. Nuts, it hurts now."

I've explained that drugs that treat anything or have any activity are called agonists. Drugs that block or reverse the action of agonists, like naloxone blocks morphine, or omeprazole (*Prilosec*) blocks acid reflux and stops heartburn, are the opposite of agonists, they are antagonists.

But here's the rub, there is a new class of drugs being developed called Biased Agonists. They are designed specifically to treat severe pain, but without the side effects of morphine, such as constipation and respiratory depression, or the big ones, tolerance, and physical dependence. Similarly, the buzz would be absent.

If that sounds like it's complicated, or contradictory, let me try to make sense of it.

I've explained that opiate receptors are on the outside of the membranes that surround our cells, mainly our brains' nerve cells.

All our cells are surrounded by membranes, only plants have cells surrounded by what are called cell walls.

Think of the receptors arrayed on your cells like your taste buds are diagrammed to be arrayed on your tongue. Opiates or opioids bind to those opiate receptors on our brain cells, known as neurons, and opioids tell those neurons, "Don't report it hurts." or maybe the opioids take away the pain, so the neurons have nothing to report, just as our taste buds tell us to say, "This cracker is both salty and sweet."

Another analogy may explain receptors more simply. Think of an electrical socket on your wall. For years, we could buy plastic plugs that fit in those sockets to keep our children from sticking bobby pins into the sockets. Those plastic plugs are antagonists. Your lamp plug is your agonist. Plug it in and your lamp goes on. Offer a nerve cell an agonist and it goes on.

Receptors communicate with our neurons through another chemical they trigger, one inside the cells that have been called *second messengers*. A fellow named Earl W. Sutherland who was a professor at Vanderbilt University in Nashville discovered the first of those second messengers and he was awarded the 1971 Nobel Prize for his discovery.

I began graduate school in the pharmacology department at Boston University School of Medicine that same year just as my mentor had begun studying the second messenger theory. Sadly, Dr. Sutherland died in 1974, a year before I finished my graduate studies.

But his discovery begins to sort out the physiology of our nervous system. We have opiate receptors because we make neurotransmitters whose job it is to transmit information by being released by one cell and to fit into the receptors on the next one. Those neurotransmitters transmit information by fitting in to receptors on the surface of the next nerve cell.

When those neurotransmitters, or the drugs we manufacture, fit into those receptors on the nearby cells, the second messenger inside the cells activate and tell the cell what to do next as it is programmed to do by genes and evolution. It's like entering a lawyer's office, walking up to the receptionist and announcing your appointment. The receptionist checks your appointment and announces to the lawyer that you've arrived. That makes you, by analogy, the neurotransmitter and makes the receptionist your second messenger.

I've already explained that drugs fit the same receptors into which the neurotransmitters ordinarily fit.

Let's complicate our lawyer's office model. What if you walked into a lawyer's office and there were two receptionists? You told the nearest one that you made an appointment online. The first receptionist pointed at the other receptionist and responded, "See her, she handles online appointments." Forgive my apparent sexism, receptionists are women in my model.

Suddenly, your arrival at your lawyer's office and your visit had to be scheduled by one of two receptionists. Remember, I said the receptionist was the office's second messenger.

It's the same with nerve cells, except there aren't two second messengers, there are at least six. That means one neurotransmitter can bind to its receptors on at least six types or anatomical locations on cells and that variety gives that single neurotransmitter more than one function. The neurotransmitter and drugs that bind to the same receptor can have six different functions, all based on what kind of cell the second messengers govern. Nature takes its time, and it is complex.

So, biased agonists are being designed to activate only one of multiple second messengers and the second messenger the agonist activates is the one that would take away pain but not the other second messengers that make us feel good, stimulate us to take the next dose or force us to increase the next dose.

The drugs are biased to have one activity, analgesia, without the nasty side effects.

Mebias' business plan is to develop drugs that stop pain but don't make users feel good or stimulate users to take the next dose. In other words, to take the abuse liability out of pain relief. My exploratory research turned up two other companies with similar business plans, *Nektar* and *Velicept Therapeutics*.

So, I sent Jim Barrett a text that included my effort to build on our meeting,

> *I'll call you Monday. I'd like to meet one or more Mebias executive and interview him or her for my book. Let me know if that works.*
>
> *Thanks & enjoy the weekend.*

His response arrived on a Monday night:

> *I spoke with them on our call today and they wanted me to say they are overcommitted and dealing with several rather pressing matters - usual things like*

funding, progression of projects, board meetings, etc. They requested revisiting sometime in April. Hopefully, this is okay and that, in the meantime, you can include features and aspects of biased signaling in your drafts.

I told him, "I'll be back in touch, thanks."

In mid-April, I revisited the issue and Jim responded, "For various reasons, the *MEBIAS* management does not wish to participate in the interview/discussion. I am sorry but they are not movable…"

I'm not one who takes "No" for an answer, so I contacted *Velicept Therapeutics* through their website.

They didn't even respond.

I've also contacted *Tevena, Inc* because they too, advertise their focus on eliminating pain without the side effects so well-known caused by morphine and the opioids. They never got back to me either.

Mebias emphasized my personal bias. Humans shouldn't suffer pain and if you've ever chopped off a finger, crushed one in a door or had surgeons filet you by sawing through your breastbone, as I have, you'll know what I'm talking about. People who take opioids are in pain, and if they end up addicted to their drug of choice, they are medical patients and deserve to be treated as such. They are not criminals to be arrested, they are victims to be helped.

What about criminal behavior, such as addicts who deal drugs to make enough money to support their habits? Their addictions are medical problems and should take precedence over their illegal activity. If they willingly enter treatment programs, and they agree to testify against their dealers, all criminal charges against them should be dropped.

Lastly, what role do you and I as non-drug users and non-drug dealers have? Write your congressional representatives and tell him or her of your priority.

Drugs, not politics.

Public Health is not political.

Purdue Frederick

In 1892, John Purdue Gray and George Frederick Bingham, two medical doctors, founded a little pharmaceutical company on Manhattan's Lower East Side. They combined their middle names and came up with the corporate name, *Purdue Frederick*. The company manufactured and sold patent medicines, including laxatives and other home remedies.

Sixty years later, in 1952, three brothers, Arthur, Raymond and Mortimer Sackler, bought *Purdue Frederick* from Gray and Bingham and moved the company headquarters to Yonkers, NY, the first town across the border from the Bronx. The three Sacklers, all psychiatrists, transformed it into a generic company, with a goal to sell drugs that out-lived their patents. Then they moved the headquarters to Stamford, CT. Arthur Sackler was the oldest brother and he died in 1987. Remaining brothers Raymond and Mortimer bought Arthur's one-third share of the company.

The Sacklers began to diversify and in 1991, they renamed the company *Purdue Pharma*.

In 2019, *Purdue University* issued the following press release:

WEST LAFAYETTE, Ind.—Purdue University is not and has never been affiliated in any way with Purdue Pharma. The pharmaceutical company was founded in Manhattan in 1892 by John Purdue Gray and George Frederick Bingham as the Purdue Frederick Company. Purdue University was founded in 1869 as Indiana's land-grant institution, named for benefactor John Purdue.

Note to Journalists: Purdue University requests that you use this paragraph in any articles on Purdue Pharma.

What happened between 1952, when the Sacklers bought *Purdue Frederick*, and 2019, when Purdue University pointedly affirmed their lack of connection with *Purdue Pharma*? Purdue University's press release, that I reproduced above, not only openly says, "…has never been affiliated in any way…" they include the *Note to Journalists*, that requests the paragraph denying affiliation with *Purdue Pharma* be included "in any articles on *Purdue Pharma*."

What happened was the Sacklers licensed oxycodone. They not only branded it *OxyContin* but they became very good, perhaps too good at selling it. Retrospectively, their marketing methods are blamed for the opioid addiction public health emergency that continues to plague us a couple of decades later.

Twenty years after Purdue licensed oxycodone, according to the U.S. Center for Disease Control, in 2016 fifty-three thousand Americans died from overdoses of an opiate drug. It's double that total now. The 2016 total was more than died from gun violence the year before and for generations, the gun industry tried to teach us that *guns don't kill people, people kill people.* It seems they were right. People use drugs to kill other people...or themselves...no irony implied.

Brothers Mortimer and Raymond Sackler, as I said, both psychiatrists from Brooklyn, did incredibly well financially from their purchase of the company. Forbes added them to their list of America's richest families in 2015 with a combined family wealth pegged at $14 billion, earned dealing drugs legally and profitably through *Purdue Pharma*. In fact, they made all that money in one generation selling *OxyContin*, their brand-name for oxycodone.

With the benefit of retrospection, *OxyContin* can be labeled as one of the most dangerous drugs ever brought to market. That danger is underscored because some people blame *OxyContin* popularity fully for the current epidemic of narcotic abuse, although narcotic abuse is a behavior influenced by all the opioids and its root extends back generations.

Oxycodone is a lot older than the Sacklers. It was first synthesized in Germany, at a University of Frankfurt laboratory in 1916, during the First World War. Those university chemists used thebaine as their starting material. Thebaine is one of the alkaloids in opium poppies, named for Thebes, the ancient Egyptian city, but it is one of the few drugs derived from poppies that is a natural stimulant, and not a natural opiate. The other, better known derivatives including morphine and codeine are depressive, in fact, each produces sleep at the right dose.

An alkaloid is a nitrogen-containing chemical made by plants, and the nitrogen in the molecule makes alkaloids basic molecules, as opposed to acids. Morphine from opium, atropine from belladonna and strychnine from the Asian flower *Strychnos* are all alkaloids. That's not to say plants don't make acids. Lemon juice, which is lemon-flavored citric acid, is one of the best examples of an acid made by a plant. Similarly, salicylic acid, the starting material for aspirin, is also a plant product. Plants are good chemists. We've learned trees pull carbon dioxide from our dirty air and make sugars from it.

Although, like morphine and codeine, thebaine is a pure product from poppy plants, it lacks any analgesic effects. As I said, thebaine is known for its ability to stimulate users, and it produces nausea and vomiting and not much more. Talented chemists changed its structure giving it morphine-like activity and it was launched by *E. Merck*, Darmstadt, the German *Merck* that remained after the U.S. *Merck* Company, that became *Merck, Sharp & Dohme*, was separated from the parent German company as part of World War 1 reparations. Oxycodone wasn't introduced into the U.S. market until 1939, two years before the Japanese military bombed Hawaii's Pearl Harbor, dragging the U.S. into World War II. Even Hitler himself was *jammed up*, modern parlance for taking oxycodone, reported by none other than Hitler's personal physician. It also underscores the idea that opioid abuse is not a new phenomenon.

DuPont Pharmaceuticals first marketed oxycodone in the U.S. way back in 1950, as *Percodan*, which was a formulation of oxycodone combined with aspirin. *Percocet*, oxycodone combined with acetaminophen, known as paracetamol in Europe, and brand-named *Tylenol* in the U.S. wasn't introduced to the market until the U.S. Bicentennial year, 1976. *Percocet* is still one of the most widely prescribed analgesics in the U.S., in part because people believe the amount of oxycodone in either *Percocet,* or even *Percodan*, isn't enough to interest addicts. Perhaps they were right, but on the other hand, together those drugs earned the street name *Percs* and it's safe to conclude that drugs of abuse earn street names, whereas aspirin is still aspirin, no matter whether in the street or the home.

The market changed in 1996, when *Purdue Pharma* released its *OxyContin* brand of oxycodone in the U.S. market. They formulated it to be a timed-release tablet, that is the suffix 'Contin', from continuous, so if a user swallowed one tablet, a small amount of the medicine at a time would be released continuously into the user's bloodstream removing the danger of abuse or overdose. That's what the company claimed, at least although street use proved them wrong.

Abusers quickly learned to defeat the timed-release mechanisms by crushing the tablets, or rubbing off the pill's coating. The also simply dissolved it in water. All three manipulations bypassed the timed-release mechanism completely, so users could take the entire oxycodone dose all at once. Taking big doses of an opiate repetitively is not only drug abuse behavior, but also the dosing pattern that leads reliably to addiction. Oxycodone even has its own street name, *Oxycotton.*

Recreational *OxyContin* use spread even as *Purdue Pharma* continued pitching the drug to doctors, hospitals, clinics, and pain treatment centers and maintained a public corporate position that their extended-release formulation was less prone to abuse than other formulations.

That was made clear on May 29, 2018, when the *NY Times* ran an article written by Barry Meier and titled, *Origins of an Epidemic: Purdue Pharma Knew Its*

Opioids Were Widely Abused. Meier reported, "a copy of a confidential Justice Department report shows that Federal prosecutors investigating the company found that *Purdue Pharma* knew about significant abuse of *OxyContin* in the first years after the drug's introduction in 1996 and concealed that information." So even though they maintained that their *OxyContin* formulation was less prone to abuse than other formulations, they were also privy to reports of abuse. There is deceit in that corporate behavior and our first tendency is to blame the company.

As I've stressed though, I think there is enough blame to go around. In the USA, FDA should have stepped up with increased or tightened regulation, physician groups should have exercised more responsibility about writing prescriptions. Drug companies must begin to understand and behave as though they are a branch of our health-care apparatus and not just incorporated mechanisms to increase earnings. It's easy to blame drug companies because they have money.

All that duplicity was under the metaphoric nose of the FDA that maintains oxycodone as a Schedule 2 drug in the U.S. Once again, the government defines Schedule 2 drugs as having *a high potential for abuse, with use potentially leading to severe psychological or physical dependence. These drugs are also considered dangerous.*

It has become an ongoing scandal in the U.S. that a Schedule 2 drug the manufacturer knew was abused, is viewed as triggering the opiate epidemic in the U.S. and is still on the market. Mortimer and Raymond Sackler both lived to ripe old ages, and in their lifetimes, they funded museums, exhibits and entire hospital and museum wings and never faced any legal responsibility for their suspension of morality and essentially triggering the opioid epidemic. *Purdue Pharma* even reformulated *OxyContin* to make it more difficult to abuse and relaunched that new formulation in August 2010. Yet that effort triggered a piece in *MedicalExpress.com* by Notre Dame's Patrick Gibbons entitled, *Study Links Rising Heroin Deaths to 2010 OxyContin Reformulation*. Purdue Pharma's new formulation didn't quite have the effect the company suggested it might. It didn't decrease abuse; it simply may have pushed users from prescription *OxyContin* to street heroin. That latter point is emphasized in the excellent article, *The Family That Built an Empire of Pain*, by Patrick Radden Keefe and appeared in the *New Yorker*, October 30, 2017.

What are we to do as a society other than mutter, "What a shame! They shouldn't have gotten away with it."

Dr. Nora Volkow, head of the U.S. *National Institute on Drug Abuse*, that's the NIDA branch of the *National Institutes of Health* (NIH), wrote in her blog in 2017, "the opioid crisis has affected some of the poorest regions of the country, such as Appalachia, and that people living in poverty are especially at risk for addiction and its consequences..."

She went on to conclude that the rise in drug addiction in poorer areas suggests it is the lack of economic opportunity that drives the drug use, and not the drug companies, ruthless drug dealers or weak personalities. It's tempting to read that and conclude we need to write our congressman or –woman and say, "You need to pass a law prohibiting American companies from moving manufacturing abroad because that produces unemployment." Another theory of addiction says that unemployment and poverty lead to isolation and it's the isolation that leads to drug use.

Those two factors also ignore the socio-economic changes in this new millennium. The U.S. is not a manufacturing economy any longer, we're spoken of as entering a *Gig Economy* with *UBER* and *Lyft* drivers the best examples of independent "Gig" workers who earn a living offering their services hourly or daily on behalf of a big company, but without being employees of the big company. Of course, that doesn't answer the question, did the jobs move abroad first, or did they move abroad because of other changes in our society? Even giant General Electric sold its locomotive manufacturing division, but of course, they also folded their financial services division, so their example proves nothing. There are changes afoot in our economy and some of those changes may drive segments of our population to despair. Maybe despair is what people exhibit when they feel isolated, and I said drug abuse is one way people deal with that isolation.

There is also a universal conclusion that opioid medicines are over-prescribed and that fits another sociological fact, opioid addiction in increasing among older folks yet death from overdose is increasing among younger folks. That supports the idea that older folks get their medication by prescription and younger folks get it from street dealers. That idea can be tested but I haven't seen such a study yet.

So, manufacturers historically pushed opioids too much, physicians overprescribed them, older folks requested them, younger folks bought them on the street or from prescription users. The result was an epidemic of addiction that shows no end.

Where do we begin?

We can't restrict market availability of opiates because there are people who function every day, for example, cancer patients, only because opiates control their pain. The number of physicians who overprescribe opiates decreases every day and there are headlines that show some physicians being arrested for their prescribing practices. Scheduling drugs does place some limits on their use, but perhaps the way Americans receive prescriptions could use some help too.

Opiate prescribing laws are administered state by state although opioid drugs receive scheduled status at the Federal level, from the US government. I contend prescribing opiates; indeed, all scheduled drugs should be Federally regulated, overseen by NIDA, FDA, or DEA, and not administered by each state. I don't think centralizing opioid drug prescriptions, indeed all scheduled drugs, would threaten States Rights advocates.

Secondly, prescriptions of scheduled drugs could more easily be tracked nationally if those prescriptions were centralized. For example, I shouldn't be able to fill an *OxyContin* prescription at my local *CVS* in the morning, travel to a *Walgreen's* half a mile away that afternoon and fill the same prescription. That is tightening up, although I live on the border of New York. I can still take my prescription across the state line to fill it. Centralizing prescriptions could make that impossible, tightening up the supply of prescription opioids.

Prescriptions for scheduled drug should never again be written on paper and handed to the patient. They should be electronic, over a secure Federal network, targeting a drug store or prescription delivery service designated by the patient with data entered by the prescribing healthcare professional. The Federal network should restrict refills for the duration of the dispensed drug's prescription. A 30-day prescription should have no refills for 29 days. No more walking down the street to fill the same prescription twice in one day, or a day before running out. We must gain control of our distribution network and that is why I emphasize centralizing it. It's a step toward socializing medicine, but I'm not advocating complete socialization, only opioids and other scheduled drugs.

Then we must accept completely that opiate addiction is a disease first, and not a crime needing trial and imprisonment. That idea began with the 1961 invention of naloxone (*Narcan*) and continues today with clinics that treat opiate addiction with the World War 2 era invention, methadone (*Dolophine*), buprenorphine (*Suboxone*), naltrexone (*Revia*) and others. The goal of acute treatment is to prevent withdrawal symptoms, universally described as unpleasant and the reason all addicts continue to take their drug of choice, to avoid that dreaded withdrawal…and seek the buzz.

Naloxone is about to become more widely available because it is so valuable in saving lives due to overdose. The *NY Times* reported on Jan 8, 2022, *New York Plans to Install 'Vending Machines' With Anti-Overdose Drugs*. New York City reports they will begin to place the units in selected neighborhoods by the end of 2022. Ambulances and emergency rooms routinely stock naloxone, especially the clever dosage form that sprays the drug up a patient's nose, invaluable if the patient is unconscious.

Then treatment experts agree, addicts also need long-term care, either group therapy or individual therapy, but the patients all need to talk. Human interaction is the best cure for isolation.

We all need to talk! Let's continue discussing opiates, their appropriate role in society and society's appropriate role in and responsibility for their regulation. We do a good job nationally regulating explosives, selling pistols, rating movies, and dating milk, but we must step up to public responsibility and step into the twenty-first century by doing a better job regulating opiates. We all have to pitch in.

We're still arguing over behaviors we don't tolerate from our children. I turned up a good example.

The U.S. House of Representatives Committee on Oversight and Reform issued a press release on Nov. 5, 2021, entitled, *Oversight Committee Launches Investigation into McKinsey & Company's Consulting Practices, Conflicts of Interest.* The *N.Y. Times* reporters Walt Bogdanich and Michael Forsythe issued an article the same day entitled; *Congress is Investigating McKinsey Over Its Role in the Opioid Crisis.* The date of their article suggests they received the press release the same day it issued.

I added both publications to my growing list of references and then *Googled* "Conflict of Interest."

Google defined conflict of interest as, "a situation in which the concerns or aims of two different parties are incompatible." *Google* also provided a second definition, "a situation in which a person is in a position to derive personal benefit from actions or decisions made in their official capacity."

How did McKinsey, a global-sized consulting company, run afoul of Congress and end up lumped in with a role in the opioid crisis?

They provided consulting advice to the F.D.A. while, at the same time, providing consulting advice to *Purdue*

Pharma, the folks who are blamed for fueling the epidemic of opioid addiction. Added to that obvious conflict of interest, McKinsey manages a hedge fund named, *McKinsey Investment Office Partners* (MIO), whose goal is to grow the savings of McKinsey employees…by investing in the very companies for which McKinsey consults.

Oops. That sounds roughly parallel to a sports team playing offense and defense at the same time, or worse, Dad telling a child, "It's Mom's fault, she taught you wrong," while at the same time Mom telling the child, "Don't listen to Dad he doesn't have your best interest at heart."

That example of consulting FDA while consulting *Purdue Pharma* not only leads to conflict, but it also violates *Google*'s definition above that says, "the aims of two parties are incompatible."

It's also another reason I'm single after two failed marriages.

It's inconceivable to me that McKinsey can consult to F.D.A. and *Purdue Pharma* simultaneously. Not only are the "aims of the two parties incompatible," they're opposite. F.D.A.'s job is to regulate the pharmaceutical industry and thereby, protect public health. They're mandated to do so by our government. *Purdue Pharma*'s job is to manufacture and sell drugs at a profit. Morality and ethics are apparently somebody else's responsibility.

Furthermore, McKinsey employees depend on their internal investment company, MIO, to grow their money by investing in, among others, the pharmaceutical industry.

The House Committee on Oversight and Reform press release went on to say, "Even today, McKinsey could benefit from its $573 million settlement with the states for its role in the opioid epidemic because of MIO's indirect ownership stakes in companies that provide treatment for substance use disorders.

The Committee's chairwoman Representative Carolyn B. Maloney wrote the following scathing summary in the press release, "MIO's opaque financial holdings raise the question of whether a consulting firm should be allowed to advise companies, governments and individuals while maintaining a hedge fund with financial interests related to that advice, without disclosing potential conflicts of interest."

The House Committee's press release ordered, "McKinsey produce documents and information about all these issues by November 19, 2021."

Committee hearings on McKinsey began five months later, in April 2022.

I could have predicted that McKinsey & Co managing partner Bob Sternfels would testify "that his consulting firm did not have a conflict of interest when it gave advice both to *OxyContin* manufacturer *Purdue Pharma* and the [FDA]." That testimony appeared in a *Reuters* press release issued April 27, 2022. The release also quoted Rep. Katie Porter of California who concluded, "Your scheme worked really well. McKinsey got contracts, Purdue got rich and America got addicted." Harsh, but deserved criticism.

McKinsey paid $573 million to State Attorneys General so the company wouldn't be prosecuted, while also admitting no wrongdoing. It's another case of, "here's a large check for you but I really did nothing wrong."

In a *NY Times* article by Michael Forsythe, Walt Bogdanich and Chris Hamby, published the same day as the *Reuters* press release, Rep. Ayanna Presseley of

Massachusetts had a similar sentiment. She said, "Your apologies feel empty and insincere."

Obviously, the U.S. Congress didn't buy Bob Sternfels argument that there was no conflict of interest.

The result of the hearings? FDA says it will not issue new contracts to McKinsey pending the outcome of the investigations.

It's all after the fact. Reactive, not proactive.

While *Purdue* maintained its *OxyContin* was safe, deaths from overdose of opioids in the U.S. continued to climb, year after year for a decade.

I find it hard to believe that no one in government noticed or asked publicly, "What's wrong here?"

As a society, we need to focus on ethics and morality at least as much as we focus on leveling income, availability of health care or education…or politics

Social Connections

As I write, we are almost two years into a coronavirus pandemic that makes writing a book on addictive disorders seem poorly timed. Yet, ever the academic, I ran into a scholarly publication in an *American Psychological Association* journal. It was written by Molly McCann Pineo and Rebecca M. Schwartz and entitled, *Commentary on the Coronavirus Pandemic: Anticipating a Fourth Wave in the Opioid Epidemic.*

How do opioid use and coronavirus infection relate to each other?

Surprisingly, the Covid pandemic triggered increased opioid use.

A 2018 *National Institutes of Health* Blog entry (https://teens.drugabuse.gov/blog/post/addiction-and-importance-social-connections) reads in part, "recent animal study suggests that positive social connections, like friendships and healthy family relationships, might undo some of the negative effects caused by drug use."

Humans are social creatures. We share emotions and feel more secure when other people are near us. It's an old concept, first voiced by Aristotle who is quoted as saying, "Man is by nature a social animal..."

But all animals seem to seek social interaction. Think about herds of buffalo, gaggles of geese, packs of dogs and so on, right down to ant colonies. There is strength in socialization.

Could a good relationship cut down on our drug use? Sounds too good to be true or at least it suggests a way to treat addiction.

The coronavirus pandemic taught us that we all need social connections to function best. None of us functioned well walled off in our homes with our masks in easy reach. For the last two months of school, my then eleven-year-old son attended his fifth-grade class through *Zoom*. I would set up my laptop across the table from him so we both had human interaction, but I noticed he looked away from his laptop screen more often than he looked toward it. He loathed remote teaching and what he missed most was being in class with other students, the socialization schools have built within them.

But how does this fit in with addictive disorders? The irrefutable statistic is that illicit drug use went up during the pandemic. At first glance, that seems paradoxical. If patients are sneezing and miserable from a viral infection, wouldn't they use fewer drugs?

No, they used more drugs.

If we focus on the idea that socializing cuts down on drug use, and the pandemic isolated us, then it makes sense that drug use went up during the pandemic. No social interaction made people seek drugs to make up for the loneliness.

What was it they had to make up for? In an earlier chapter, I ascribed opioid use to a desire to lose a bad feeling. I theorize that loneliness is the bad feeling many opioid users want to rid themselves of. The pandemic made us all feel lonely.

First, let's accept that we're still in the middle of an epidemic of opioid abuse. Almost 47,000 deaths from opioid overdose alone have been reported each year for 2017 and 2018, rising to 108,000 in 2022. The U.S. has lost about 150,000 people to the coronavirus pandemic and yet we lose that many opioid users every two years, and we have for decades.

The surprise is how much social contact plays in our mental health. The other surprise is how "hardwired" we are for social interaction. Think about it, the primary way we seek information from our environment is through our vision. Yet blind people seek the same social interaction the rest of us seek. They're not seeking that social interaction through their vision; they are hardwired for it.

So, death from opioid overdose is every bit as big a problem as the pandemic. It's just been overshadowed because deaths from Covid rose rapidly, peaked, and declined all in two years. Deaths from opioid overdose has climbed every year for a decade and shows no sign of letting up. Remember, for the year ending in 2022, the U.S. lost 108,000 people to opioid overdose.

Let's get together as we are wired to do. The need for social interaction is one of the reasons recovering drug addicts need aftercare. They need someone to tell them wordlessly, perhaps, "You're okay." Go to your church, synagogue, or your mosque. Go to a town meeting or your child's school PTA meeting. Organize a block meeting or resident's meeting. Join a craft club, take a course, listen to a lecture, or lead one. If you're in recovery, meet with your group. Make every effort to be with other people.

If you write, go to the library.

Primum Non Nocere

Hippocrates was right, translated he said, "Above all, do no harm."

A recent headline read, *How NJ might use $641 million in opioid settlement money*, the article written by journalist Dustin Racioppi and appeared in a local weekly paper, usually tossed onto my driveway.

Where did New Jersey get so much money?

It was part of a settlement paid by The *Johnson & Johnson* company and three drug distributors.

Why or what did the *Band-Aid* manufacturer and some distributors have to settle?

I did some digging and found a press release, dated February 3, 2022, published by an organization that calls itself North Jersey (www.northjersey.com). The headline read, *New Jersey in line for $641 million from opioids settlement, a lifeline for treatment*.

Johnson and Johnson put out a press release about two weeks later, on February 25, 2022, the headline of which read, *Johnson & Johnson Statement in Nationwide Opioid Settlement Agreement*. The company's press release had the following paragraph: "As previously announced on Jul 21, 2021, the Company will contribute up to $5 billion to the nationwide settlement, which is designed to directly support state and local efforts to make meaningful progress in addressing the opioid crisis in the United States. This settlement agreement is not an admission of any liability or wrongdoing, and the Company will continue to defend against any litigation that this final settlement agreement does not resolve. The company no longer sells prescription opioid medications in the United States as part of our ongoing efforts to focus on transformational innovation and serving unmet medical needs."

Wow! Here's $5 billion but we don't admit we did anything wrong and, "Oh, by-the-way, we don't sell opioids anymore." Forgive my cynicism, but if I'm going to write a big check to pay someone not to sue me, then it's safe to assume I did something wrong. Before lawyers begin shouting at me, I do understand that by saying they're not admitting wrongdoing, prevents former patients from suing the company for negligence.

What did J&J do wrong? They fell in behind the bandwagon leading a parade the entire pharmaceutical industry joined, namely selling prescription opioid drugs. And what's New Jersey going to do with all that money (Yes, I live in New Jersey). Quoting Racioppi's article, "New Jersey plans to invest $641 million the next two decades in harm reduction services, adding money to substance abuse programs..." I hope so because the Covid epidemic closed a lot of treatment facilities.

The idea of harm reduction is new to the U.S. but hammered home by Canada. The city of Vancouver, B.C. hosts a website entitled, *Harm Reduction – Vancouver Coastal Health* in which they stress the idea of a needle recovery program that includes a hotline to call for used needle pickup.

Back in 2003, Vancouver, British Columbia established a program they named *Insite*. It provides a place active addicts can feel safe while they take their drugs of abuse. Abusers won't be arrested; while it gets them off the street and provides someone to help them if they overdose.

It is a leading example of what Vincent Martello called Harm Reduction.

The U.S. is on the same page as the Canadians regarding Harm Reduction. The U.S. Surgeon General urges Americans carry naloxone. There's an inconsistency there though, naloxone is still a prescription drug in many states.

Vancouver's Harm Reduction strategy showed such good results that San Francisco, Philadelphia, New York and Seattle have all sent representatives to learn about Vancouver's supervised injection facility.

Europeans embraced the idea of harm reduction years ago.

Insite is in the Vancouver neighborhood that historically was the center of the local drug trade. Putting it in that neighborhood made it available to addicts. It brought the facility to them and didn't ask them to find the facility. A spokesman says, "That's part of what this place and space is about. It's not just about keeping people alive – that's the primary goal – it's about making a space where drug users are allowed to feel like people."

That's entirely consistent with the idea I introduced in my first chapter, that my addiction counselor students wanted to know if I was an addict because no one else ever paid them any attention.

Another spokesman said of *Insite*, that rates of fatal overdoses were averted annually, both in the facility and the surrounding neighborhood, in part, because addicts were less likely to share needles. The surprise though, was addicts increasingly started to seek voluntary rehab or detox and as I've pointed out, voluntarily seeking help is key to success.

How did Vancouver get ahead of the curve? Journalist Travis Lupick wrote, "Vancouver's embrace of Harm Reduction was entirely led by activists and by this community. It was an entire decade of concerted effort and activism that really brought in the government kicking and screaming at the end."

Those activists even shook loose funding from Canada's government.

It's another example of a community-based treatment program that works.

The *Canadian Drug Policy Coalition* defines harm reduction as a service that, "enhances the ability of people who use substances to have increased control over their lives and their health and allows them to take protective and proactive measures for themselves, their families and their communities." Those are all keys to good mental health.

Vancouver also offers, as part of their harm reduction services, needle exchange, supervised consumption sites and overdose prevention and response services.

It's a model system that many American cities would do well to emulate.

In fact, several American cities have introduced supervised injection centers like Vancouver's "supervised consumption." It's a controversial move, however, because there are segments of the U.S. population who believe strongly that government shouldn't be in the business of supplying drug addicts.

That's short-sighted though, because it's one thing to say, "We have to keep drug addiction from spreading." It's more realistic to accept that people are going to abuse drugs so let's be responsible for our fellow humans and lessen their risk.

Think of an image, knights in shining armor jousting. Their armor represented harm reduction. They were less likely to be hurt by their opponent's lance or sword.

Recognizing that heroin addicts are going to inject their drug and providing a safe place for them to inject is also harm reduction.

It is not throwing in the towel and admitting preventive measures don't work. It's saying, "While we work on preventive measures, we accept you've got a medical problem. We're there for your safety."

It's a huge step forward. It's bringing safety measures to the addicts instead of waiting for a few addicts to say, "Please help me."

I've referenced harm reduction in earlier chapters, but I want to emphasize it because it is overdue and it works.

Needle exchange, naloxone availability, trained addiction counselors and safety from arrest are the characteristics of harm reduction services.

Needle exchange allows addicts who inject their drugs of choice a source of sterile needles to eliminate shooting up with yesterday's, found, or shared needles.

Naloxone is what I termed a life saver and harm reduction services are a ready source of naloxone and people trained to use it.

Trained counselors not only know how to use naloxone, but they also recognize addicts who endanger themselves through overdose and they are prepared to help.

Safety from arrest finally accepts my preaching that addicts are public health victims, not criminals. Police are not outside harm reduction facilities waiting to arrest patients for using drugs.

New York City opened a harm reduction facility in 2021. Way to go!

There's a new take on the idea, bring harm reduction to the addict. House calls!

At the opposite end of New York's population is tiny Hickory, N.C. I read a guest essay in the NY Times, written by Beth Macy, entitled *The Two Simple Edicts of Successful Addiction Treatment*.

It inspired me to dig deeper and I found a website (https://www.integratedcarehickory.com). Their website was headlined "Not Treatment, Recovery." It was subtitled "Beat Depression or Anxiety, Kick the Drug or Alcohol Habit… With Our Help." How refreshing, and not judgmental.

I copied their text freely because it was so inspiring: *All treatment plans for patients incorporate medication management along with counseling, as well as other effective interventions. Due to the ravages of drugs and alcohol in our society, our services for substance use disorder focus on the Twelve-Step counseling model with medication-assisted treatment using buprenorphine-containing medications (Suboxone, Zubsolv, etc.). But we aren't your usual drug rehab or treatment center! We will work with you closely to maintain focus on your recovery and coming off of any replacement medications slowly over time. Our goal is to help you organize your life and live happy, joyous, and free OUTSIDE of our clinic.*

Harm reduction at its finest. Notice they say, "We will work with you closely…" It's what recovery requires, constant aftercare, lack of judgment, freedom from arrest and a promise of a future.

Let's use New York City's leadership and Hickory, NC's example of the right way to organize our fight against opioid addiction. It's our fight, not their fight. It is our society invaded by the metaphoric infection that is opioid addiction. Addicts are the sick members and the rest of us must help.

Let's help.

Back to companies paying penalties, a *New York Times* article, written by Jan Hoffman and published March 3, 2022, read in part, "…the Sacklers would pay as much as $6 billion to help communities address the damage from the opioid crisis." The Sackler family controlled *Purdue Pharma*, the company responsible for *OxyContin,* the drug that dominated the opioids market and led the increase in opioid sales that became a crisis.

If we look back ever further to 2021, *Teva* pharmaceuticals was found liable in a suit brought by New York state. *Teva*'s fine was to be levied later. Earlier that year, *Allergan* was also fined $200 million by New York state. *Teva* was a major supplier of generic opioid pharmaceuticals. To add insult to injury, *Teva* bought the generic line from *Allergan*.

A July 2022 article on Bloomberg reported, "Teva Pharmaceutical will Pay Over $4 Billion in Opioid Settlement." The article went on to say, "The Israeli drugmaker said Tuesday, it reached a tentative deal to pay $3 billion in cash and $1.2 billion in donated overdose-fighting drug Narcan to resolve claims."

More power to them for owning up to the problem.

I am not indicting the industry that supported me so well for a couple of decades. I am pointing out that our legal system has placed some responsibility on the pharmaceutical industry, and the industry has stepped up and paid up. I offer, they assumed responsibility for easy prescription opioid drug availability and went on with their business. Perhaps it's time for other players to assume responsibility and get on with their business.

For example, I also said earlier, the U.S. FDA could have been stricter. They are a regulatory agency, after all. We probably didn't need a dozen opioid drugs on the market in five formulations each, ranging from tablets to patches and capsules to injections. Alas, there is no admission of culpability from FDA although they have tightened controls because of the overuse. I'd like to see them more proactive than reactive and perhaps that's a subject for future debate, or leadership.

Some in our medical community let patient requests govern what they prescribed rather than rely on their medical training. It's a bit like my now thirteen-year-old saying to me while lying in his orthodontist's chair, "Dad, I don't want braces."

My response was, "I don't care. It's a medical decision and I'm saying you'll tolerate it."

Physicians need to recognize when they're being played by their patients. If patients have conditions that hurt so much the physicians decides to treat them with opioids, physicians should also treat the conditions causing the pain, if medically possible, and not just treat the pain itself.

There is a seemingly paradoxical statistic regarding deaths from opioid overdose. They have risen to more than 100,000 a year during the Covid epidemic as though one caused the other. That's even in the face of the huge payments from J&J and *Purdue Pharma* and the decreased availability of prescription opioids, both from decreased manufacturing and tighter control over prescription writing and increased awareness in the medical profession. The increase is due to illicit street drugs smuggled into the U.S.

I concluded that our population was isolated during the Covid crisis and people turned to drugs because of that isolation. Furthermore, they turned to illicit drugs because the new controls decreased prescription drug availability, and huge industry payments had only begun to appear. I have no statistics to support that conclusion, but it is sensible.

Maia Szalavitz, writing in a guest essay in the *NY Times*, asks, "When did I yield to temptation – in a fit of rage over a boyfriend's infidelity in the mid-1980s…It was relief from my dread and anxiety, and a soothing sense that I was safe, nurtured and unconditionally loved." Her essay appeared Dec. 6, 2021, and she titled it, *Opioids Feel Like Love. That's why they're deadly in Tough Times.*

It's tough to argue about wanting to feel loved and wanted.

The other factor driving the increase in deaths from overdose is the increasing presence of fentanyl and its potent derivatives first contaminating the heroin supply, and now growing to fentanyl contamination of most illicit drugs of abuse. That fentanyl was traced to China, and much of it still originates there. It continues to be smuggled into Mexico and Latin America. From there, it's smuggled into the U.S. We've still not been able to block that smuggling nor convinced the Chinese that it's not in the world's best interest to continue supplying fentanyl, or at least to interfere with its underground manufacture.

Another *NY Times* article published April 23, 2021, reports the death rate from drug overdose exceeded the death rate from coronavirus in 2020. The numbers are stunning, in San Francisco overdose deaths were triple deaths from the epidemic.

That increase was hammered home by a report in the same newspaper, by Shawn McCreesh on April 14, 2021. He wrote that he is from Hatboro, PA and, "During high school, prescription pills were as easy to abuse as a learner's permit. Our reunions take place coffin-side and often."

Similarly, treatment centers closed during the covid epidemic, further limiting opportunity, and stimulating opioid use and overdose.

The isolation brought on by the Covid epidemic causing people to turn to drugs is consistent with the idea I presented in my chapter on our endogenous opioid receptors. I said social interaction increased levels of our endogenous opioids, known as endorphins. Increasing our endorphin levels is associated with feelings of well-being. Conversely, as our endorphin levels fall, people may be motivated to keep those receptors occupied by taking opioid drugs. The Covid epidemic caused social isolation, decreasing our endorphin levels, and stimulating illicit drug use.

Scary thought.

There's a flip side to this story as detailed in a *NY Times* article also written by Maia Szalavitz, entitled, *What the Opioid Crisis Took From People in Pain.* It appeared March 7, 2022. She makes a point that with all the new controls and decreased availability of opioid prescriptions, that patients in severe pain, like cancer patients and accident victims, suffer a loss of their quality of life because their pain is untreated. She details one patient in her article who took his own life because of his untreated pain.

Hippocrates is called the Father of Medicine and he is credited with the oath new physicians are requested to take. His oath has within it, the concept of, "Above all, do no harm, *Primum non nocere.*"

Leaving a patient to suffer pain is doing harm so it leaves us with a choice that hasn't formed a middle ground yet. That choice is, prescribe opioids for patients in pain, but don't make those opioids available to people whose goal is to abuse them.

There must be a middle ground, but medical practice hasn't found it yet.

What is that middle ground?

As I said earlier, we don't need quite as many opioids on the market as we've had. That's a future pharmaceutical industry negotiation with the U.S. FDA.

Secondly, we don't need quite as many formulations as we have. Let me propose that we have two opioids for parenteral administration (by injection), two for oral administration, two for patches and perhaps one for nasal administration. That's it.

Thirdly, I propose again that opioid prescriptions be tracked nationally, not by state. Whether that's by FDA, DEA or NIH, needs to be worked out. That would provide physicians with a documented answer to the question asked of patients, "Are you prescribed this drug by any other physician or hospital?" Easy to check with a central system.

Lastly, physicians need a way to judge the sincerity of a patient's pain. For example, is the patient suffering cancer or other life-threatening disease? Is the patient an accident victim trying to get back on his or her feet? Is the patient suffering an age-related disability? These are a few examples of a situation medical professionals must agree on, define among themselves and set universal guidelines.

Those four steps might tighten up prescription opioid use without denying patients in pain the ability to lead normal lives. They don't address illicit opioid use.

What does?

Tightening up illicit fentanyl manufacture and smuggling. If indeed illicit fentanyl comes from China, surely a totalitarian government like theirs, can get control of this. It's gone on too long.

Better "sniffing" technology to root out drugs hidden in commercial shipments or in simple suitcases. I maintain that if I own technology that lets me see who is ringing my doorbell, we can certainly develop technology that alerts us if a bag of fentanyl is buried in a trailer-load of clothing.

We also have to stimulate more social interaction among American communities. What would be so wrong about monthly meetings with the general topic of avoiding drug abuse?

What's a community according to my suggestion? It could be a block organization, residents of a building or a village. A town meeting, a church or synagogue service or a high school assembly are excellent opportunities to stimulate interaction. If scouting is still alive, a monthly scout meeting dedicated to the subject. How about a high school class on the general subject of maintaining health? It goes beyond washing hands, showering regularly, wearing a mask, or wearing a condom. It goes to what one puts in his or her body.

Above All, Do No Harm!

Looking Forward

While it's important to look back, learn where we came from, and study the surprising discovery and age of opiates, the invention synthetic opioids, and their effects, it's equally important to look forward and ask a few new questions.

The first that comes to mind, "Is recovery from addiction possible?

A variation on that question might be, "How does an addict begin recovery other than by hospital admission through the Emergency Room, or commitment by a court?"

The question that nags addicts, "After recovery, can I have a full life?"

"What's the role of community in addiction recovery?"

"Is the U.S. government active in this area of public health?"

I started this side project by reaching out to my local Congressional Representative as I repeatedly advise my readers to do. I don't believe public health is political, quite the opposite, it's universal.

He is Rep. Josh Gottheimer of New Jersey's fifth Congressional district, the northern area of the state that borders New York. I wrote to his legislative aide, Cody Hollerich who obviously passed my questions up to Rep. Gottheimer.

Rep. Gottheimer responded, "Thank you for your support for increased resources to fight addiction."

I had asked about pending legislation to address addiction and rehabilitation.

He went on, "...I am proud to be an original co-sponsor of the *Excellence in Mental Health and Addiction Treatment Expansion Act*, which will expand the existing demonstration programs in the Fifth District and improve access to community mental health and addiction services."

While that was a bit of legislative fresh air, he continued with, "I was also proud to vote for the bipartisan SUPPORT for *Patients and Communities Act*, which was signed into law on October 24, 2018. This vital legislation combats the opioid epidemic by increasing access to addiction treatment."

I think it's refreshing and appropriate that my Congressional Representative is active in legislating Federal support for addiction treatment, but I want to turn from pending Federal legislation to the public health issue that asks the nitty-gritty question, how do we initiate recovery in the addict population in our community? What is a community? Can it be a university campus, a town, a city, county, or state?

When I complained to one of my readers, a psychotherapist, "I'm looking for inspiration to write the last half of my book. I want to refocus my direction on the future and away from discussing history and technology."

His response was immediate as he nodded 'yes' saying slowly, "I think I can help." In the past after I asked him to read pieces of my work, he encouraged me to continue without marking up my manuscript. I didn't know what to expect this time and he surprised me.

He stood and walked to his desk, wrote something, and walked back to me saying, "I know an ex-addict who would like you to interview him for your book." He handed me a small piece of notebook paper with a telephone number written on it.

I was floored because I'd been trying to interview recovering addicts for this book since I'd begun writing and I referenced a few early attempts, including contacting *Narcotics Anonymous*, a few hospitals and several rehabilitation organizations and I always emphasized, "I make every effort to keep identities private."

I never had much success, because I was told some variety of, "It's a violation of anonymity to identify a recovering addict." *Narcotics Anonymous* told me to contact someone in the rehabilitation industry. I also complained to the psychotherapist about inspiration solely to complain, not to seek help.

I looked at my psychotherapist friend and asked, "He's waiting for my call?"

"I've told him about you," he responded, "He's looking forward to talking to you."

I gladly noted the contact number the psychotherapist gave me, called it that afternoon, left a message, and received a return call the same day. We made an appointment to meet the following week on a Thursday evening, at my home. I chose Thursday because it's the day my young son visits his mom, giving me a little more freedom to focus on my book.

So, I'm writing this on a Friday because last night, as I write, was our meeting. My doorbell rang at the time we'd scheduled, and I welcomed a middle-aged, successful man I'll call Peter and invited him into my living room. He had called me at least three times to confirm our meeting suggesting he was as anxious or enthusiastic to speak to me as I was pleased to speak to him. He was about six feet tall, dressed casually with his hair the same shade of grey I as mine. He had more pepper and less salt than I have although he wore no beard, so my white beard wasn't echoed. We spent an interesting hour chatting about addiction and its effects.

Peter is an excellent example of the new face of opioid addiction, an educated, successful middle-class worker, who had a good life before addiction and was able to re-establish it in recovery.

Opioid addiction is one of the few patterns of modern life that spread from disadvantaged classes of people toward more advantaged populations. Other features of our society spread in the opposite direction, from upper and middle classes to less advantaged people. Think about the spread of smartphones although their spread, and that of technology generally, is also held back by affordability. Whether we like to admit it or not, our society is stratified. Opioid use and abuse differentiate even more when we consider, a century ago it was a practice of upper-class women. More recently, it was rampant in less advantaged populations and from there, spread into the middle and upper class. Interestingly, most users say the same thing when asked why they started, "I needed to get rid of the stress in my life. I felt lousy."

Among less advantaged, that stress is often, financial need. Conversely, more advantaged folks cite business, or familial need producing stress, demonstrating that as we push forward in our careers or families, stress follows us although we allow it to arise from sources other than money.

Peter summarized that idea when I asked him how he started using opioid drugs. "I started so I could function better with all the stress in my life."

"Stress from what?" I asked.

"I'm referring to business stress, not financial pressure," he explained. I learned he runs his own corporation; he has a business partner, and his marriage also survived his addiction.

I reiterated my philosophy about opioid addiction, addicts are patients, not criminals and they need help, not prison.

"So, did you start with a prescription for an injury and then find it impossible to quit." I continued our conversation with the too-common reason people end up hooked on opioids, they begin treating pain from injury, surgery, or illness with a prescription before they find themselves saying something like, "The drug got its hooks into me, now I need it to function."

Peter responded, "Oh, it was a prescription, it just wasn't prescribed to me. I bought my drug supply from a friend who had too many prescribed to him. My friend was prescribed 270 *OxyContin* pills a month, too many for anyone, so I bought his extra pills," he explained. "I was spending $200 a day at my peak, over $70,000 a year."

It's a further example of excessive prescriptions for *OxyContin* and consistent with news reports about the drug's history.

"That's nine pills a day for pain," I said, shaking my own head. "That's how the manufacturer's reputation was damaged. Too many opioids were prescribed," I muttered.

As I worked on this chapter, I found a website from *Carlisle Medical*, a nation-wide managed care site that advertises "Carlisle Medical has over 40 years' experience as a leader in the workers' compensation industry." Their website led me to a white paper they published entitled,

Fighting the War on Opioids in the Workers' Compensation Industry, and their on-line cover page has the following questions, "Is this too much? Should a claimant need this much pain medication for this long?"

It fits so well with Peter's neighbor's experience of receiving a prescription for 270 opiate drug pills a month and is so on target with news' reports of excessive opioid drug prescriptions.

I continued Peter's interview by asking, "What made you keep taking it?"

He responded, as if by reflex, "When I didn't take it, I felt sick."

"Withdrawal," I said quietly, nodding knowingly.

"Probably," he agreed after some thought.

"Can you describe what you mean by feeling sick for me?" A recurring question that has fascinated me for years.

"I would begin sweating and breathing fast and those responses wouldn't stop. I couldn't concentrate." He described his withdrawal as though it happened the day before. It seemed to be a bitter memory he carried near the surface and that memory wanted to escape from where he'd trapped it.

"When did you start taking drugs?" I asked.

"About a year, maybe a year or a year and a half ago. I really can't remember." He shrugged.

"What made you start?" I asked, rephrasing my original thought.

"Stress wouldn't let me function," he said, crinkling his face and repeating the idea.

"Did you take anything other than *OxyContin*?" I asked.

"Yeah, fentanyl." He looked at me directly but without emotion, as though admitting fentanyl use was shameful.

"Dangerous stuff," I observed. "Are you OK, on time," I added, changing the subject briefly, because when he arrived, he mentioned he had another appointment after our chat.

"I'm good 'til eight," he said glancing at his own watch. It was about 7:30 I noted on my own watch.

"How did you recover?" I refocused our discussion of his story.

Without pause and with a direct look, he said, "I couldn't function anymore without looking frequently at my watch to see when I was due to take my next dose. I remember one event I attended with my wife. I sat there sweating as I waited 'til I could sneak away and take a pill."

"How long were you in recovery?" I continued the theme.

"A week. They put me on *Suboxone* (buprenorphine and naloxone) and now I go to counseling two days a week, as well as group therapy and a psychotherapist. I still take the *Suboxone*."

Suboxone is used to treat opioid addiction because the two drugs in it, buprenorphine, and naloxone, are designed to sit on our opiate receptors but not produce any effects and if patients should try to take their usual drugs of abuse, *Suboxone* blocks the effects of any other opioid the addict takes so there's no buzz. It also prevents withdrawal sickness because the two drugs in *Suboxone* sit on our opioid receptors, but they don't stimulate the receptors to respond.

"Sounds like you've managed to shake loose the drug's grip on you. Congratulations." I said and then continued," Have you seen any change in your life's goal?"

"That's an interesting question," he responded and paused to think about his answer. "I feel enormous gratitude that I didn't overdose and kill myself." He paused again before adding, "That's not really a goal, is it?"

"No, but it's a nice emotion. Very positive."

"Besides," he continued," I always try to help people stay off the stuff, but I've learned if you tell someone to stop, they won't listen anyway, so I gave up trying to tell people that."

"But you seem to have a strong memory of bottoming out and now you're recovering. Your recovery would be a good model for people to emulate. Maybe, looking at how you recovered, or that there's a good life waiting for you after recovery should be your message," I understood his desire to help others yet not know how to help. It's why I write now and I taught addicts as I described in my first chapter.

He didn't respond but his face displayed his interest. That also ended our interview. He looked at his watch and said, "I've got to get going. Would you like me to come by again?"

"Definitely, thanks. I need to write this up and then we'll talk."

I followed him outside and noticed something. Coincidentally, he and I drive the same car although mine's blue and his is silver.

We met again two weeks later, and I was curious about his aftercare. I wanted to understand it and I asked him to summarize it for me.

"On Mondays, I have a Men's Group, on Tuesdays we have relapse prevention, Wednesdays I see my psychotherapist and Thursdays I meet with my counselor."

His aftercare totaled to four-days a week, but I asked, "Do you find yourself reaching out for more social contact since you left rehab?" He corrected me because what I had called rehab, he called detox.

"No, I spend too many hours on the phone running my business, so I don't have a lot of time or energy to meet new people at night."

I was curious about his social life, so I asked, "Have any of your relationships changed since Detox?"

He was thoughtful before responding, "I still find myself wanting to help people all the time, but I don't know how to."

I also inquired if his marriage had changed since Detox.

He sat up straight as he smiled, again nodding slowly. He said, maintaining his smile, "Yes, I'm not hiding anything from her anymore. When I walk in the front door, I leave my ego outside and concentrate on her."

I was curious if he found himself making long-range plans and he responded, "Yes, we're trying to buy a beach house, maybe even move into it one day. It's funny when I was using, I never made any plans beyond my next dose."

We chatted for almost an hour and our subject touched on new emotions. He said, "I feel hungry all the time and I don't know what's driving that hunger." Then he added, "Sometimes, when I close my eyes, I see a vision of someplace, so I open my eyes to see where I am and realize wherever it was, it was just a vision. Sometimes it's scary and I wonder if it's my brain's way of looking at my past or forcing me to look at it."

Peter is also a good example of the idea that seeking help voluntarily is more likely to lead to successful recovery, as opposed to being committed by a court proceeding for treatment or any therapy forced on a drug user by family or civil commitment.

I'll even go so far as to suggest that civil, family or court commitment to involuntary rehab is just one step away from labeling addicts as criminals and jailing them for drug use. Involuntary commitment seems punitive and not likely to become voluntary, whereas voluntarily seeking rehabilitation, is not punitive. It is based on a patient seeking to help him- or herself and is motivated by desire and not by society forcing it on the addict. Peter is an excellent example of voluntarily seeking detox and then returning successfully to his social, family and business life.

The U.S. has a way to go before we give all patients the opportunity to seek help voluntarily. Thirty-seven states have laws that mandate rehabilitation. While I said mandated rehab is a step beyond labeling addiction as criminal behavior, it's still not giving addicts the same freedom of choice given someone with a sprained ankle, a bellyache, or a toothache. We're all tempted to say, "Make that phone call and find someone to help."

But it also calls up our economic reality. Involuntary commitment frequently involves the state, county or local government paying for detox or rehabilitation. Voluntary rehabilitation usually involves paying for the process out of the patient's pocket or the patient's insurance company pays it as a claim. That almost guarantees a discordant separation between patients with financial means and disadvantaged patients. Disadvantaged patients need a way to pay for rehabilitation and in many states, that almost pushes them into involuntary commitment.

I'll turn now from individual recovery to community programs. Ulster County, NY is an example of a community-based effort to eliminate those economic barriers to rehabilitation by making it available to county residents at no cost. It is also the county in New York with the best rehabilitation record of every other county in the state and a good social or political model for other states, counties, or municipalities to follow.

The numbers support that:

In 2018, the County had 280 recorded cases of overdose of which 20% were fatal.

By 2021, the County had cut their rate of overdose almost in half, to 142 recorded cases, although the rate of overdose deaths was similar at 19.2 %.

That overdose death rate is one reason New York City, fifty miles south of Ulster County, has elected to install Naloxone vending machines at selected spots around the city. It's an attitude that shouts the reality, "We know you're going to take drugs, but we don't want you to die because of your habit." But that's New York City. It has a city-wide attitude of acceptance, problem solving and progress, all at full speed.

I met with Ulster County's Director of Community Health Relations who, by his own admission, "Came into public health by the backdoor. I was in public communication when I came into this office." His name is Vincent Martello and he is a white-haired 70-year-old professional with a resonantly deep baritone voice that held my attention. I think I expected him to sing his response, his voice was so resonant.

After he shared with me the success his programs had in reducing overdoses, he added, "The only way to treat overdose successfully, is to treat the whole person and not just the overdose."

I responded, "Voluntary rehab is far more successful than forced rehab either by the courts or family." I had done my homework.

He nodded and went on to describe what he has labeled a *Strategic Framework for Confronting the Opioid Epidemic*. He said, "There are four pillars of that Framework, Reduce Supply, Reduce Demand, Reduce Harm and Improve the System of Treatment, Care and Recovery," and he handed me a printout summarizing those pillars.

Reducing Supply includes combatting trafficking, engaging providers, educating patients, taking back their drugs and disposing of them safely and securing medications legally.

Reducing Demand includes increasing Risk Awareness, Engaging High-Risk populations including schools, primarily preventing initial use and prevention-based communications campaigns. It uses education to keep people from starting drugs. Increasing social interaction by any means is the goal.

His focus on Reducing Harm includes responding rapidly to medical emergencies including widespread access to *Narcan* (naloxone), around the clock peer support, fentanyl test strips and identifying high-risk persons. I didn't know fentanyl test strips were an available technology, so I did some research.

They're like pH paper that most of us have seen in high school chemistry. Fentanyl test strips detect fentanyl in street drugs, or in urine, and the test takes five minutes. Powdered drugs must first be dissolved in water and the test strip dipped into the mixture. They're accurate in detecting fentanyl but they won't report how much is there. I even found a website published by the city of New York with instructions for testing drugs with the strips. It's a realistic harm-reduction strategy. (https://www1.nyc.gov/assets/doh/downloads/pdf/basas/fentanyl-test-strips-brochure.pdf).

Vincent's last Framework pillar ended with, "Establish a high-risk mitigation support team to support individuals from overdose through multi-year path to recovery.

It's an ambitious program and has begun to serve as a model to other counties who've watched Ulster County's success rate. He emphasized that recovery is multi-year and will likely have to be initiated more than once because the relapse rate is high. "That's to be expected," he added.

A month after our meeting, I corresponded with him electronically, and Vincent shared with me his most recent statistics. County-wide, for the twelve months ending mid-year 2022, Ulster County New York reported 232 overdoses, almost half from heroin, with 20 deaths due to overdose. That calculates to 11.6 % of overdoses were fatal and is about half the death rate he reported earlier. He also showed, for the same period, 37 % of all overdoses didn't receive naloxone, suggesting naloxone in overdose is just what I call it, a lifesaver.

Together, Peter's experience and Vincent's leadership are good models to follow.

Peter voluntarily entered rehab, maintained his marriage, and re-entered his career. It serves as validation of the idea that individuals can drive their own recovery if they have a good support structure in place.

Ulster County's model serves any community with a framework for confronting the opioid epidemic and validation that the overdose rate and deaths due to overdose can be managed successfully as a public health model that makes aftercare available population-wide. Communities can learn from Ulster County's example.

Let's conquer this public health menace together.

As European journalist Johan Hari has opined, "The opposite of addiction is not sobriety; the opposite of addiction is human connection."

Let's all connect.

Is the Future Psychedelic?

Continuing our attempt to glance the future, or at least the direction we're headed, drug treatment for addiction may become possible with psychedelic agents, although scientifically they are still unproven and not licensed.

As an aside, I came through the 60s psychedelic era unscathed. I kept my thinking straight and never took LSD, psilocybin, mescaline, or any agent other than single malt scotch with no ice or a few beers. Back in high school, my friend Bob and I used to drive into Staten Island to buy beer because New York allowed alcohol purchase at eighteen back then, while most of the rest of the country set twenty-one as the age one could buy alcohol. In winter, we'd hide the beer in the snow until my parents went out on a Saturday night. Then we partied, albeit underaged, from New Jersey's standards. After college, I indulged too much, but I never even smoked dope, let alone dropped acid.

But our future may include psychedelic drugs as part of addiction therapy and not just as recreational drugs for an occasional trip on Saturday night.

First some background.

The word *psychedelic* was coined by psychiatrist Humphry Osmond in 1956, at a meeting of the *New York Academy of Sciences*. Osmond defined psychedelic as "mind manifesting," and it lent its name to the psychedelic era of the sixties. We can understand what he meant by mind manifesting if we think of our minds displaying images or sounds unexpectedly, sometimes reversing them. We imagine seeing sounds and hearing colors. We've labeled those unexpected images and sounds hallucinations. They're one of the characteristic effects of psychedelic drugs.

In June 2017, a review article appeared in the medical journal *Neurotherapeutic,* written by two Johns Hopkins University doctors, Dr. Mathew W. Johnson, and Roland R. Griffiths, who entitled their article, *Potential Therapeutic Effects of Psilocybin.*

Psilocybin is a natural product. It is a psychedelic agent extracted from *Peyote* cactus. That source was discovered by Native Americans generations before, suggesting they had their own psychedelic era for centuries. That little cactus is native to Northern Mexico and is found through the Southwestern U.S. In fact, the word *Peyote* seems to derive from the *Aztec* word for the cactus.

LSD is another common psychedelic agent, although by contrast it is not natural, but rather a product of modern chemistry labs. Psilocybin and LSD seem to work the same way, by binding to some subpopulation of our brain's serotonin receptors. Serotonin is one of our brain neurotransmitters, chemicals whose job it is to facilitate how our nerves communicate with each other. It's much as I described for how our brains use enkephalins to influence our mood and our bodies use insulin to reduce circulating sugar. Both enkephalins and insulin work by binding to their specific receptors, where they activate a second messenger, and that second messenger "tells" the cells to which they bind, "Do your thing," that is, respond to the binding as physiology has evolved them to do. Serotonin works similarly, in that it binds to several receptors in our brains, and the cells that have serotonin receptors respond as evolution taught them.

But does psylocibin help SUD (substance use disorder) patients conquer their opioid addiction by binding to those serotonin receptors?

The article written by those two Johns Hopkins doctors concluded, "Psilocybin in the treatment of addiction is currently at an earlier stage of research..." Their article discussed how the drug seemed to help people quit smoking or drinking, both examples of addiction, but I write about opioids, and they were saying, "There are not enough data yet."

So, even though there's no proof yet, many scientists are dusting off old medical literature to review as they struggle to secure research funding so they can test psilocybin with modern, balanced, and valid experimental designs, the results of which would stand muster with the U.S. FDA or the *European Medicines Agency* or EMA.

The big experimental design factor missing so far, is known as "blinding," meaning neither the patient nor the doctor knows whether the patient is receiving a drug or a placebo, or if the patient received the drug, what dose did he or she receive, and in what order? That way, "bias" is removed from the study design, bias ascribed to the physicians evaluating the patient's response or the patients reporting their responses unobjectively, that is, with preconceived notions of what they expect. Blinded studies allow patients to respond objectively, that is without any preconceived idea what to expect because they don't know what they received, and those responses can be compared with normal, accepted statistical methods. When neither the doctor nor the patient knows what the patient is taking, it's called a "double blind" study and it's the holy grail of medical research.

Dr. Jerry Avorn, in his excellent book, *Powerful Medicines, The Benefits, Risks, and Costs of Prescription Drugs*, summed up that idea of biased drug studies, "The problem is this: patients who end up taking a certain drug can differ in important and unmeasured ways from those who do not, even if they appear to be otherwise similar." In other words, biased results.

The path led by Drs. Johnson and Griffiths nevertheless resulted in a coup, of sorts, Johns Hopkins became the first U.S. university granted regulatory approval for a research center they named, *Center for Psychedelic and Consciousness Research*, part of their esteemed medical school.

A 2019 article by reporter Barbara Sprunt, from radio station WAMU, a public radio station owned by American University, emphasized that in the US, psychedelic research isn't widely funded by the Federal government because psychedelic drugs have always been classified as Schedule One. The drugs cannot legally be prescribed to patients because they are defined as having no medical use and are therefore illegal at the Federal level. That outflanks states' rights and what I referred to as unlicensed. Hence, the center had to wait for private funding to initiate their work because they couldn't apply to the government for research funding to study drugs already defined by our government as dangerous. She concluded her article by saying, "The center at Johns Hopkins plans to test psylocibin in opioid addiction among other psychiatric disorders."

They did. I found an article also published in 2019, in the medical journal *Frontiers in Psychiatry*, written by a group of six collaborating scientists including Drs. Griffith and Johnson, whom I mentioned above. The new article concluded, that after an online, retrospective survey, "The majority of respondents reported meeting DSM-5 [*Diagnostic and Statistical Manual*] criteria for severe SUD before their psychedelic experience, whereas in the time since their psychedelic experience, the majority no longer met criteria for any SUD." In other words, the psylocibin seemed to have worked because after their psychedelic experience, the patients did not show addictive behaviors. However, as I explained above, there was no blinding in their study. The patients knew what drug they were given and so did the doctors because retrospective studies use older, unblinded data. Both the patients and the doctors knew what the patients were given. Additionally, it was retrospective, meaning a re-examination of an old study, although they did use an objective measure, the *Diagnostic and Statistical Manual* applied modern, quantitative criteria.

While their online survey doesn't meet modern standards of statistical validity, it still suggests strongly that psychedelics are useful in treating SUD, or at least, they need more study because seeming to work is not the same as proving they work. I did see a valid, so-called, as I explained above, "double-blind" study of psylocibin in depression and anxiety in cancer patients. That suggests further that the drug has promise treating SUD patients, but we must be patient, er, wait a while longer for the right study.

I'll continue to watch for medical reports of its use in SUD patients, but as I write, psylocibin treatment of addiction is still preliminary, yet early data suggests it may prove useful.

Standby, because proving utility will require tens of millions of dollars funding, no matter what the source. As I said above, however, Federal funding in the U.S. for psychedelic research in treating SUD, isn't available because psychedelic drugs are Schedule I, so the field depends heavily on funding from private, non-governmental sources and I appeal to those sources to begin funding medical research to the same extent they fund arts, artists, and TV documentaries.

It would be a good opportunity for foundations or private or corporate wealth to contribute heavily to public health research for a condition that is killing 100,000 Americans annually.

It's another case of Federal laws needing to catch up with the need to stimulate progress in addiction research to provide public health guidance by providing answers to outstanding questions.

Let's go Congress too, less politics and more public health! Health is not political.

There are yet a few more drugs in this class of psychedelics. The first is known as ibogaine and it's a natural product also found in plants.

Any time we begin to think extracting drugs from plants it is a throwback to an earlier era during which humans ate nuts and berries they picked from trees. But it's not a throwback because every time I tear open an envelope of the *Truvia* brand of sugar-free stevia sweetener, I read the envelope and it has printed on it, "Calorie-free sweetener from the Stevia leaf." We're still eating natural products and will continue to do so. I have my eye on the wild strawberries ripening in my garden as I write.

Ibogaine grows in Central Africa and a 120 years ago, French explorers brought it back to Europe. It's rumored that the U.S. CIA used ibogaine in the 1950s although it would be up to an imaginative spy novelist to suggest to us how they used it or for what.

It's not approved in the U.S. as a medicine although it is approved in some other countries including Mexico. Nevertheless, it is being studied in the U.S. in Miami although I found few details on how they managed to get away with it.

It is proposed to block withdrawal in opioid users although it awaits either corporate or governmental sponsorship of real clinical trials that include that blinding design I mentioned earlier.

Like ibogaine, "toad venom," or 5-MeO-DMT is another psychedelic agent derived from both the Sonoran Desert Toad and some plants. It's yet another drug that blocks some of our serotonin receptors and begs to be studied in recovering addicts (SUD patients) because it has been shown to be effective in depression and anxiety patients.

Treating addiction with psychedelic agents is leading to, at least state-by-state efforts to decriminalize psychedelics to allow physicians the freedom to test the drugs in addicts without fear of violating the law. The states, however, don't have the wealth the Federal government displays.

The N.Y. Times ran an article on May 21, 2022, entitled, *I Want to Reset My Brain: Female Veterans Turn to Psychedelic Therapy*. Since psychedelics remain Schedule I drugs in the U.S., female veterans check themselves into a Mexican clinic. The reporter, Ernesto Lonono writes, "Having turned to experimental therapies to treat post-traumatic stress disorder, traumatic brain injuries, addiction and depression, many former military members have become effusive advocates for a wider embrace of psychedelics."

The Mexican clinic is named *The Mission Within*, run by Dr. Martin Polanco, and it treats U.S. veterans with psychedelics outside the reach of U.S. law enforcement.

It's interesting and overdue, that he treats women.

While decreasing our opium supply seems not to fit in a chapter on psychedelic agents, most scientists agree that decreasing the U.S. heroin supply must be a simultaneous focus during withdrawal treatment.

Briefly turning our attention to decreasing the opium supply, estimates two decades ago, ranked Afghanistan first, producing as much as 90% of the world's opium. Since then, it's been banned by the Taliban, which decreased Afghanistan's production although it didn't wipe it out. There were other decreases in the region too. Turkey abolished opium production as far back as 1972 and it drove up Afghanistan's production.

Similarly, Iran banned poppy production in 1979 which also kicked poppy production to Afghanistan.

We must look further East, toward China.

And now we turn to another possibility, could marijuana also be useful in preventing opioid drug addiction? Again, it's probably not proven, but a lot of talented people study it.

I found a 2014 medical article in the academic journal *JAMA Intern Med*. The chief author, Dr. Marcus A. Bachhuber is based in Philadelphia and there were two other scientists listed as co-authors. Their contention was deaths from opioid overdose continue to rise, driven by increases in chronic pain prescriptions, and chronic pain is a major indication for medical cannabis, could changes in laws governing access to medical marijuana reverse the increasing rate of deaths due to opioid overdose.

As I explained earlier, theirs was a retrospective study of state cannabis laws and if they correlate with decreasing opioid overdose deaths.

Nevertheless, their conclusions stimulated, so to speak, increased interest in using medical marijuana prescriptions to prevent opioid use. Like many retrospective studies, however, Dr. Bachhuber and his colleagues concluded, "Although the present study provides evidence that medical cannabis laws are associated with reductions in opioid analgesic mortality on a population level, proposed mechanisms for this association are speculative and rely on indirect evidence."

In other words, Dr. Bachhuber told us yet again, correlation does not mean causation.

An article five years later written by four authors headed by Dr. Chelsea Shover, appeared in another medical journal, *PNAS*, or *Proceedings of the National Academy of Science*. They opened their article on a hopeful note, "Medical cannabis has been touted as a solution to the U.S. opioid overdose crisis since Bachhuber..." Dr. Shover and her colleagues stated without reservation, "We find it unlikely that medical cannabis – used by about 2.5 % of the U.S. population- has exerted large conflicting effects on opioid overdose mortality."

Repeating the principle, correlation does not mean causation, cannabis doesn't prevent opioid overdose. Dr. Shover said it best, "Cannabinoids have demonstrated therapeutic benefits but reducing population-level opioid overdose mortality does not appear to be among them."

The research results are tantalizing, and scientists will continue to study these drugs.

Lastly, we can't forget the drug known as ecstasy, or more formally as MDMA. I found an article in another medical journal (*Current Drug Abuse Reviews*), published back in 2013 by three New York–based academicians entitled, "Can MDMA Play a Role in the Treatment of Substance Abuse?" While they admit MDMA works by a mechanism that's different from other psychedelic agents, it may be useful in substance abuse because it works in PTSD patients.

That's a stretch because, while PTSD is one causative factor in drug abuse, it's not the only cause. Nevertheless, if academic labs can shake loose research funding from either the government, private foundations, or other sources, we can learn the answer to the intriguing questions surrounding psychedelic treatment of SUD patients.

Strategic Change in a Post-Covid World

Maya Angelou said, "If you don't like something, change it. If you can't change it, change your attitude." Changing our attitude applies to substance abuse treatment in a post-Covid world.

There was an attempt to gain control of the spread of opioid dependency that was either interrupted or subsumed by the recent Covid pandemic. As the Covid pandemic winds down, we must refocus on opioids because the deathrate from opioid overdose continues to rise.

The U.S. DEA, an arm of the Department of Justice is charged with regulating availability of opioid drugs. The Justice Department's Inspector General accused the DEA of being slow to respond to the growing epidemic of death from opioid overdose. That sounds like a regulatory agency being proactive until we look at when the Inspector General said that. It was back in 2019, making it reactive and, as it turned out, too late. Covid was just beginning to spread.

The *U.S. Council of the Inspectors General on Integrity and Efficiency* issued a report in October 2019 entitled *Combatting the Opioid Crisis: Role of the Inspector General Community*.

Two months later, in December 2019, the Covid virus began its spread from China to the extent that 2020 will always be remembered as the year we stopped eating in restaurants, public schools taught their students online, we were required to wear masks and we all stayed home.

It's no wonder that just as the U.S. Government began its focus on combatting Covid, the deathrate from opioids continued to rise, exceeding 100,000 overdose deaths a year.

I gained a new respect for the U.S. government's efforts to fight death from opioid overdose. That Inspector General (IG) report opened with the statement, "The United States is in the middle of a severe opioid crisis. More than 70,000 people died from drug overdoses in 2017…"

The IG report continued with, "The 2017 report of the President's Commission on Combatting Drug Addiction and the opioid crisis made 56 separate recommendations to improve the government response to opioid abuse…" I couldn't get a copy of those 56 recommendations, but it was gratifying to learn the U.S. Federal Government was focused on opioid overdose deaths but, like the rest of Western Culture, was badly distracted by a more acute pandemic.

Medical researcher Hugo Lopez-Pelayo and his colleagues wrote a piece in *BMC Medicine*, published July 2020 entitled, *The post-COVID era: challenges in the treatment of substance use disorder (SUD) after the pandemic*." He went on to say, "The COVID-19 pandemic offers a unique opportunity to reshape and update addiction treatment networks."

My Congressional Representative Josh Gottheimer reported in 2020, "GOTTHEIMER SOUNDS ALARM ON OPIOID EPIDEMIC WORSENING DURING COVID PANDEMIC."

He went on to stress his observation, "The COVID-19 crisis is only making the opioid epidemic worse, with added economic stress, anxiety, and depression brought on by being home alone contributing to an increase in substance and alcohol use."

My thanks to Cody Hollerich, Rep. Gottheimer's Senior Legislative Assistant for responding to and keeping up our dialogue.

The pandemic decreased the number of treatment options for drug abusers in the U.S., in part, because Covid patients dominated Emergency Rooms for nearly a year and a half. Surprisingly, Lopez-Pelayo writes from Portugal and proposes that modern SUD treatment incorporate telemedicine and digital solutions, suggesting again, SUD is not uniquely an American problem, but a global problem. He goes on to explain to us that technology can provide continuity of care because it removes the need for patients to visit a facility. It essentially brings the facility to the patient. Of course, he also admits that digital technology can exclude groups of people who don't have access to the technology because they can't afford it and, if our Covid pandemic is any lesson, we are not adept when we try on-line learning. We don't like it either. It's a skill that will require some practice before we're comfortable and easy-going with it, should the need arise again in the future. Other than reading the daily paper online and researching this book, I can't read online for pleasure either.

That *BMC Medicine* article also stressed that SUD treatment should be modernized because the last time it was updated was forty years ago. It's long overdue, although we have enough experience to know what worked and what needs updating. I'm a big fan of applying technology to solving the problem of increasing contact between SUD patients and care providers.

Those renewed treatment systems should be based on what the article termed "the seven pillars, including:

Telemedicine and digital solutions

Home hospitalizations

Consultation/liaison psychiatric and addiction services

Harm reduction facilities

Person-centered care

Paid work to improve quality of life in people with SUD

Integrated addiction care.

Let's think about those pillars one at a time.

Telemedicine, or what I refer to as our "*Zoom*" era, of necessity will result in less biological testing such as needing blood, urine, saliva, or tissue samples, but it is also a less-expensive way to provide contact with a broader population. The fly-in-the-ointment to which I keep returning, is digital communication technologies like laptops or personal devices like iPads aren't universally affordable so until access to the technology is universal, the communication technology offers is limited to people of means. I'm against adopting solutions to problems if those solutions further stratify our population.

Home hospitalization is almost a contradiction, but we can interpret it as receiving daily contact with a caregiver at home, again, largely provided by technology. Continuity of contact is key in addiction therapy.

Hospital admissions due to substance abuse is a valid way to begin, but it's only a beginning. It will help patients through withdrawal but provides no structure for ongoing care. It does save lives though. We were cut off from hospital visits during Covid though, because Covid patients, as I said, dominated hospital ERs for a year.

Harm reduction includes abstinence but that is a goal, not a starting point. A stable home is a good starting place, but it focuses on providing a home for patients who don't have one, or at least a stable living environment. I don't know if the U.S. homeless population abuses opioids, but it's a good idea to keep people from living in the streets. The big fish in harm reduction is providing safe locations for injection, staffed with life-saving experts and naloxone. They are not welcomed universally. "Not in my backyard," is too common.

My Congressional Representative, Josh Gottheimer put out another newsletter that seized on harm reduction in a select population. His piece was headlined, "GOTTHEIMER INTRODUCES BIPARTISAN, BICAMERAL BILL TO COMBAT STUDENT ATHLETE OPIOID ADDICTION. CREATES FEDERAL YOUTH EDUCATIONAL & TRAINING GRANT PROGRAM ON PREVENTION"

He explained his bill as follows, "…March 28, 2022, U.S. Congressman Josh Gottheimer (NJ-5) announced that he is introducing bipartisan, bicameral legislation known as the *Student Athlete Opioid Prevention Act*. The legislation will create a federal grant program through the *Substance Abuse and Mental Health Services Administration* (SAMHSA), a branch of the U.S. Department of Health and Human Services, to invest in educational and training programs at the youth, high school, and collegiate levels on the misuse of opioids and other substances commonly used in pain management or injury recovery by students and student athletes." It's a repeating story that pain management leads to addictive behavior that, in turn, leads to addiction.

It's gratifying to me to see the U.S. Congress attack drug addiction head-on with a rational, albeit an athletic-focused, program, based on treatment models that have been proven by others.

Person-centered care undoubtedly began with recognizing and accepting substance abusers as patients first, not criminals and continued with the concept that patients fare better if they say, "I want to get clean," as opposed to treatment mandated by the court or family, as in "get clean or else."

Paid work is a goal of many businesses in a post-covid world. Even grocery stores don't open all checkout aisles any longer, in part because they can't find workers to staff them. How many of us, in our post-Covid world, have gone into a grocery store and searched vainly for someone to ask, "What aisle is toilet paper?" Worker shortage is a problem facing our economy and may need major overhaul in how we do business. As I write, the U.S. went through a shortage of baby formula, blamed, in part, on monopolization by the baby formula industry. There aren't enough manufacturers, so if the biggest producer goes off-line because of supply-line or quality control issues, it creates a shortage. What's next, we'll begin ordering everything from *Amazon* and the idea of shopping in person will become archaic?

Integrated addiction care is both triggered, and victimized, by the huge increase in hospital populations driven by Covid infection. For SUD patients, we must move toward an integrated treatment model that includes early detection, intervention, and specialty hospital care. There are facilities available, and they advertise insurance is accepted, but in-patient care during recovery is still not universally available, and if a patient doesn't have insurance, those facilities, like universal access to technology, remain out of reach. It's the conundrum of stratification, someone invents a machine or a technique to treat a population but it's use is limited by availability of funds.

The global nature of death from overdose is also evidenced by Scotland. That small nation has a deathrate from overdose that rivals the U.S. A 2019 *NY Times* article quoted Andrew McAuley, Senior Research Fellow on substance abuse at Glasgow Caledonian University, "I remember when the headlines said "One Death Every Day From Drugs…That seems like glory days. We're four times that now." There is a small distinction between Scotland's deathrate from overdose and that of the U.S., Scotland's victims tend to be older than those in the U.S. We'd have to theorize why that is.

Despite the global nature of the disease of opioid addiction, it's time to re-examine how we treat a disease that is killing 100,000 Americans a year and will continue to do so until we figure out how to make treatment universal, determine what works and what doesn't, and provide some measure of continued care to prevent relapse. After all, the two highest death rates from overdose in the U.S. include West Virginia…and the Bronx Burrough of NYC. And the main culprit in the NYC deathrate, fentanyl.

As Thoreau said, long before Maya Angelou told us to change our attitude, "Things do not change; we change."

Let's change together. If we all work toward the same goal, we can accomplish it. Congress is on it…finally. The Justice Department was on it and the Executive Branch needs to get on it.

How are We Doing?

I've said deaths from overdose reached 108,000 in 2022, a deathrate that makes me gulp! Nearly three hundred people a day dying from opioid overdose, in a country that can send a rocket to Mars, just to say we did it, but, at the same time, not yet developed a program to prevent drug abuse.

Are we making any progress at all?

Our progress is uneven, especially so because the death rate continues to rise and it varies to groups and races within the population.

Caitlin White, managing editor of *Health City* at *Boston Medical Center*, quotes Dr. Marc LaRochelle, a specialist in addiction at that Medical Center, "In the last three years, it looks like the brakes had been put on for white population, but other groups are still going up."

We are still wrestling with uneven opportunity in the American population, although now availability of medical care is the culprit. Minority groups used to be denied access to equal education, many schools were "White Only." Similarly, minority groups didn't have open access to housing or transportation. Quite simply, they were denied equal life opportunity solely based on how they looked, what language they spoke at home, or more recently, what gender they declared. Now it's access to medical care for a public health issue and not denial of education or public transportation.

She goes on to write, "Since 2016, however, the story has changed. While white fatalities have decreased through 2019, opioid overdose deaths among Black Americans – particularly Black men – are accelerating." When we consider that deaths from opioid overdose almost doubled in the last decade, it's clear where that rate is rising, among Black Americans. I submit that some of that is due to accessibility of health care, although I can't rule out a genetic basis for the deathrate.

She goes on to write, "U.S. experts mark the start of the current opioid epidemic with the rise of prescription opioid analgesic use for non-medical reasons, particularly *OxyContin*, in the 1990s." She jumps ahead to the following conclusion, "But in 2013, the third wave of the epidemic posted a steep rise in synthetic opioid overdose deaths due mainly to fentanyl."

I've covered that in earlier chapters.

The AMA (*American Medical Association*) reports in their 2021 white paper, *Overdose Epidemic, Physicians' progress toward ending the nation's overdose epidemic,* "Opioid prescriptions decrease for 10th consecutive year, but deaths continue to increase, it's time to change course." They go on to report, "Physicians and other health care professionals have reduced opioid prescribing in every state for 10 consecutive years. They have increased the use of state prescription drug monitoring programs (PDMPs) in every state for the past five years. Despite these efforts, drug-related mortality continues to rise."

The U.S. death rate from opioid overdose doesn't seem to respond to decreased prescription writing. The AMA recognized that and recommended, "Treating the nation's drug overdose and death epidemic demands a far more proactive and coordinated approach focused on evidence-based public health solutions."

Opioid addiction is still a public health crisis with no end in sight.

Psychiatrists have their own take on the crisis. The *American Journal of Psychiatry*, in a 2017 academic article by Dr. George E. Woody, recommends yet another treatment for maintaining recovering addicts free of opioids. It's a drug called naltrexone and he wrote, "Interest in naltrexone increased when studies found that it prevented relapse to alcohol dependence."

Alcohol dependence is a separate subject from opioids, but naltrexone has been formulated as an extended-release injection and, unlike methadone, it is an antagonist, not an agonist. Similarly, buprenorphine is a partial agonist, useful in addiction rehabilitation.

Naltrexone use must wait for SUD patients to be clear of their drug of abuse as well as clear of either naloxone or buprenorphine, but it's success rate, if the rules are followed, is increasing. Waiting for patients to be clear of their drugs of abuse is vital because as an antagonist, naltrexone precipitates withdrawal. Methadone doesn't because it is an agonist although, as I've pointed out, that also leads methadone into the community as a drug of abuse.

So, how are we doing?

It's clear the deathrate due to overdose continues to increase. That suggests we have an ongoing problem even if some states report fewer cases of addiction, people still die from opioids.

That scary deathrate is unevenly suffered by the U.S. population. Opioid addiction may have escaped epidemic proportions in disadvantaged populations, but the deathrate stayed behind. As I said, a larger percentage of Black addicts die than white addicts die.

We need pain drugs that don't produce addictive behavior. People still suffer debilitating injuries and cancer, and those patients need their severe pain treated. Today, only opioids can treat their pain. But opioids cause addiction if the drugs leave the hospital settings those patients require, they enter the addiction trade.

Health insurance companies must universally treat addiction therapy without question. If addicts voluntarily enter treatment, and they have insurance, the treatment, including their aftercare, should be covered. Those two statements need to be debated and legislation passed before they could take effect, but insurance should do the job it's advertised to do.

There is enough evidence that recovery from opioid addiction is fraught with relapse and that relapse rate decreases with adequate aftercare. We need government-supported aftercare for addiction treatment. It would be a good idea if such aftercare followed a new model, patients who have insurance or adequate means would pay their fair share. Patients without insurance or adequate means, should have that aftercare provided by a government agency. Perhaps Medicaid could make it one of their paid services. For example, what if companies that sell or distribute opioids, pay a tax on those opioids, or pay for a license to manufacture and sell them. That money could flow toward aftercare. My suggestion needs an economist's input to be effective but Medicaid is an existing solution.

Finally, we need to acknowledge and make law that opioid addicts are not criminals, they're patients in a public health emergency. Addicts imprisoned for using drugs should have their convictions vacated, or at least re-examined.

We can learn from our recent Covid pandemic, public health emergencies need public response and cooperation. We all wore masks for two years so we know we can do it. Those masks prevented wearers from catching Covid and prevented them from spreading the virus.

Opioid drugs kill people, disrupt families, clog our courts, decrease productivity, and interrupt lives.

Let's all help. Start by learning what you can and if your community wants to build a safe injection site near your home, don't fight them. Patients enter those harm reduction sites because they can inject in a clean environment and someone is there to save them if they overdose. The experience has the added benefit that patients ask to enter rehab or detox at a high rate. It's motivational.

Is Opium Really That Easy to Grow?

Apparently not, at least in my hands.

Below is a photograph of my bottle of poppy seeds I keep in my kitchen because I like to make my own bread and I add the seeds to some of my doughs. I store the poppy seeds with all the other spices in a cabinet above my stove. There's nothing special about them and I've had them two years, as I write. I don't even know if they're *Papaver somniferum,* the opium poppy, or some other species because it's not written on the jar.

I thought I'd try to sprout them, so I planted them in peat pots filled with potting soil, watered them regularly and when I returned home after a weekend away, they'd sprouted, as I show in the photo below. I've kept the pots on the sill of a basement window facing west. Sprouting took about 10 days to the stage pictured. I water them three times a week. The sprouts are tiny and I'm anxious to see them grow.

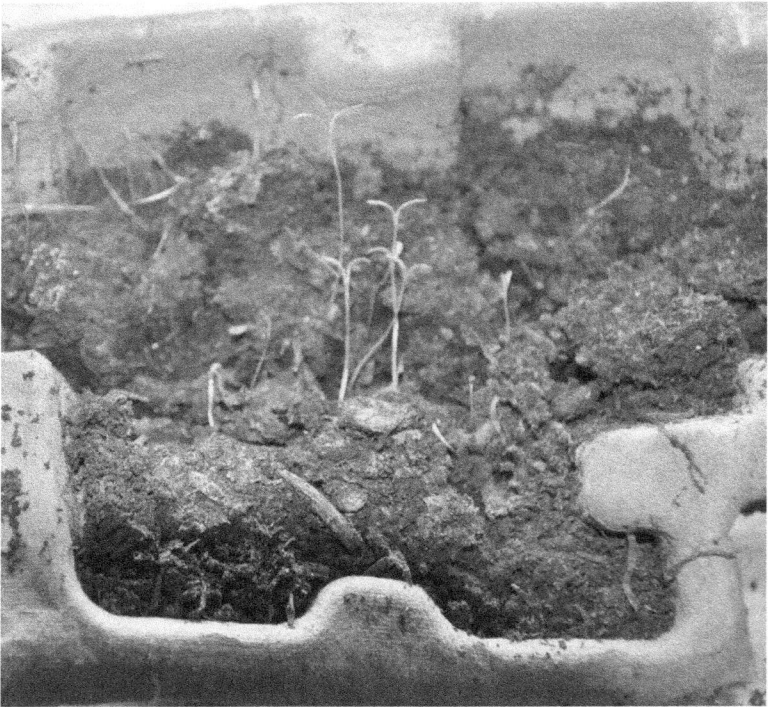

I mentioned this project to Cathey, the woman I've dated for three years, but she was cautionary and didn't share my enthusiasm, "Don't raise them to harvest the raw opium, that's against the law and you could get in trouble."

"That's why I planted my kitchen supply, there should be very little opium in the plants," I defended my activity.

Stay tuned.

Three weeks have passed and the sprouts in the photo above, have matured to have second and third leaves, but the stems of the plants are still tiny and not sturdy enough to support the plants.

I've selected an outdoor area that I hope to clean of weeds in a few weeks so I can plant the poppy seedlings.

Alas, I put the seedlings outside although still in their peat pots. They died. So, while poppies seeds are easy to sprout, in my hands, they wouldn't grow past their secondary leaves. Oh well!

Never one to give up easily, I bought poppy seeds from a garden shop and repeated planting them in peat pots, but this time I kept them outside where they could get sun and be watered daily. They sprouted, grew their second leaves and I planted the peat pots.

The seedlings died.

I quit.

SAMADHI

Vincent Martello, Ulster County, NY's Director of Community Relations, whom I introduced in an earlier chapter, copied me on an email that said, "I just wanted to put the 2 of you together." The other copy went to David McNamara which gave me David's email address, so I sent David an email message entitled, "Vincent Martello said we should touch base."

David responded. He's managing director of a treatment facility in Kingston, NY with the head-scratching name *SAMADHI*, a reference to Buddhist philosophy.

We made an appointment to meet, and he offered to give me a tour his facility. I drove up this morning and just returned. It was 140-mile round trip, but worth it.

David is a fifty-eight-year-old dedicated opponent of drug addiction. He has the same salt 'n' pepper hair as I have but his beard is a grey goatee. He's built the way I was before I lost 60 pounds. I admit when I was fifty-eight, my hair was still brown, although my weight was two-eighteen. Now, I'm 160.

He gave me the brief tour he had offered, and I saw their art room, group therapy room, a couple of offices and then we retired to his office. He'd taken off his boots and was walking around in stocking feet.

I asked, "Can I keep my sneakers on?"

"Of course," he answered as he ushered me into his sparsely furnished office. There was a picture of Buddha over his desk.

I asked, "How did you happen to use art as part of your therapy program?"

"It seems to work; we gather everyone into the room and tell them draw what you feel."

"Maybe it's just offering a social environment that was missing in their lives." I said, surprised by art as a treatment modality.

He began talking, another person eager to share his story. I asked how he launched his treatment organization.

"I spent my savings on it," he began. He then went on to describe how Vincent Martello gave him his first contract that allowed SAMADHI to grant "32 scholarships to train recovery coaches, and many of those people are still with us."

"Oh, so you do collaborate with Vincent," I muttered and went on, "What's the structure of SAMADHI"

"We're a non-profit," he said.

Then he continued, "I came to this from the film industry where I produced documentaries and commercials. I also encountered Gabor Maté and the Robin Hood Foundation."

One of the background books I read as I researched this book was Maté's *In the Realm of Hungry Ghosts*. I said, "I mentioned to my agent that my book subject would make a good documentary".

He looked at me and answered, "I haven't finished the film yet." Hmmm!

The Robin Hood Foundation is a New York based charity that addresses problems due to poverty.

David also described how his *SAMADHI* organization works with prisons and jails in Ulster County, NY, not far from Vincent's headquarters. He visits jails, speaks with prisoners, and added, "Many of them visit us here when they're released because we're the only friendly face they'd encountered while locked up. We offer to train them as counselors"

I circled back to ask how he started with Eastern philosophy and his answer surprised me. "I started with martial arts, specifically Japanese karate. From there I embraced Buddhism in a purer form." His explanation suggested to me that his spirit fighting addiction comes from the same spirit that drove him to study Japanese karate.

Our meeting lasted but an hour because I had to get home to make lunch for my youngest. Single parenting, especially in my seventies, drives my schedule. I wouldn't have it any other way.

Oh, and for those of us with little exposure to Asian philosophy, *SAMADHI* means, among other things, "the last stage in direct knowledge through identification."

It clearly helps in rehabilitating addicts.

Meeting with David also triggered a question that rattled in my brain, so I wrote to Vincent, "I drove up and met with David this morning for about an hour. Can you bring me up to speed on the extent to which you contract with David for aftercare? It's an interesting public/private arrangement that could serve as a model elsewhere."

Vincent's response arrived within minutes, "Great. I'm glad that you guys were able to get together. Over the years, we have helped fund Samadhi's activities through a series of grants, including Opioid Data to Action (CDC via NYSDOH), *Columbia University's Healing Communities Study* and a few other smaller ones. The average grant is around 100K.and are federally sourced funds that pass down to State Health Departments and then to counties. Samadhi also received substantial private funding from the *Novo Foundation* (Peter Buffet who lives in UC [Ulster County]." NYSDOH is New York State Department of Health.

It is indeed an interesting Public/Private arrangement that could be copied, and I'm pleased to be able to share it with my readers.

Who's in Charge?

I've proposed centralizing opioid prescriptions at the Federal level to make it easier to limit and track those prescriptions. It would also avoid the duplication exhibited by our current State-run systems, avoid too many doses distributed to one patient as I described for Peter's neighbor who was prescribed nine doses a day.

Who would manage the system I proposed?

The U.S. doesn't have a lot of experience with centralized health management. Aside from the Veteran's Administration (VA), Indian Health Service (IHS), Medicaid and Military medicine, the U.S. has tended to distribute responsibility to manage medical care to the States.

That leaves open the question of who manages or coordinates the U.S. agencies that regulate drugs in our current system. We still provide healthcare as private enterprise although we're beginning to see mergers of joint practices, those practices conglomerating and Amazon buying practices.

The U.S. management agencies include, on the regulatory side, the Drug Enforcement Administration (DEA), the National Institute on Drug Abuse (NIDA), the Food and Drug Administration (FDA), the Substance Abuse and Mental Health Services Administration (SAMHSA) and the Centers for Disease Prevention and Control (CDC)?

They not only have different missions and separate management structures, but they are managed by different Cabinet heads. Borrowing a management concept from my team management responsibilities when I worked in the industry, it's important to bring in people with different areas of expertise if there's a leadership in place, even if leadership's only visible task is to carry the Board of Directors direction back to the workers and worker's ideas and concerns back to senior management. Team management with different disciplines worked well.

NIDA, for example is part of the *National Institutes of Health* (NIH) and is charged with basic research. They describe their mandate on their website, "Our mission is to advance science on the causes and consequences of drug use and addiction and to apply that knowledge to improve individual and public health." The NIH has risen to the top of world science leadership and productivity and as a destination for advanced training in the medical sciences. They are an important forward-looking asset in our struggle against opioid drug abuse.

The DEA, by contrast, is a law enforcement agency. Their website says their mandate is, "enforcing the controlled substances laws and regulations of the United States. While this includes investigating criminals and drug gangs that traffic in illegal drugs." DEA is a *Justice Department* agency, and they have the power to investigate, infiltrate and arrest subjects. When I pulled up their website, I found a press release they issued in June 2022, *Ten Arrested in Southeastern Massachusetts Fentanyl Trafficking Conspiracy 14.9 kilograms of suspected fentanyl seized.*

They're very good at meeting the goals of their mandate. If more illicit drugs get through our defenses, it's not because the DEA is asleep at the wheel. They're outnumbered on the world stage and they need more technology.

FDA defines its mission as, "Protecting the public health by assuring that food (except for meat from livestock, poultry and some egg products which are regulated by the U.S. *Department of Agriculture*) are safe, wholesome, sanitary and properly labeled; ensuring that human and veterinary drugs, and vaccines and other biological products and medical devices intended for human use are safe and effective." Their European counterpart is the *European Medicines Agency*. Without a license from FDA (or EMA) to market a drug or vaccine, pharmaceutical companies cannot introduce either agent to market. To apply for a license, a pharmaceutical company must conduct clinical trials to show their new product works in human patients and is safe for them. Additionally, their manufacturing practices have to pass muster and labeling must be approved.

The *Substance Abuse and Mental Health Services Administration* (SAMHSA) defines their mission with the statement, "We lead public health efforts to advance the behavioral health of the nation." Their job is critical, especially after my discussion of the frequent comorbidity of mental illness and opioid use. I've also repeated that I think of opioid abuse and mental health as falling into line on a continuum.

The *Centers for Medicare and Medicaid Services* is a Federal and State supported program. *Medicaid* pays for some health services for low-income citizens while *Medicare* pays for some health services for citizens over age 65. Some senior citizens receive benefits from both *Medicare* and *Medicaid*. Those programs are a step toward socialized medicine with Federal and State governments paying, in part, for health care although, in the U.S., healthcare is still sought and provided privately. I doubt the U.S. will ever socialize medical care for the entire population, but individual agencies such as Medicare and Medicaid, Indian Health, Military Medicine and the VA, all borrow heavily from central government control and begin to resemble socialization.

Lastly, who would manage international issues as police agencies continue to fight drug smuggling? Some of that falls to *DEA* but smuggling also involves the *Border Patrol* and *Customs*. Those law enforcement agencies are administered by a single program entitled *U.S. Customs and Border Patrol*. Given that most of the illicit drugs coming into the U.S., flow from Mexico, however, the drug sources are in China and other Asian countries.

Most seizures today are of drugs hidden in hidden in cavities in car bodies or ride on semi-trailers whose main loads are agricultural products or manufactured goods. Given the potency of heroin and fentanyl, the amounts smuggled in each load is small. As I demonstrated in my calculations, though, because the dosages are small, a few kilos go a long way. It's another place I stress that technology can help.

I've already discussed my idea to harness technology to "sniff" drugs hidden in other shipments. I'm not a technologist but as I said, if my cell phone can receive emails sent by my my car's software, building a machine that can find hidden drugs shouldn't be far behind.

I propose, the U.S. needs an addiction leader to manage our drug program centrally.

What's an addiction leader? I've thought about it as a Presidential Commission, usually defined as a Presidentially appointed task force charged with investigating something. While it would have the power to manage the process, it doesn't sound like the right choice for leading a restructured system to manage prescriptions and fight illicit opioids.

So, I propose the U.S. establish a cabinet-level appointment whose mandate would be to coordinate activity among and between the five agencies I listed above, maintain the database of opioid drug prescriptions I proposed earlier, meet and work with the *United Nations Office on Drugs and Crime* (UNODC) and interact with the *European Medicines Agency*.

It's a big task, but it's largely a reorganization that doesn't require anything more than real estate to house the new management, and computer power to restructure how the U.S. tracks and stores opioid prescriptions. That must be created from scratch, after somehow restructuring the law to transfer authority for opioid prescriptions from the states.

I'm not a politician but I recognize that latter task is a political challenge. I repeat my mantra, public health is not political, it should be universal.

My proposed restructuring doesn't solve the problem of illicit or street drugs flowing in, nor does it address the need for available aftercare for the entire population. But it's a start.

It would bring the five agencies that now report to different cabinet-level departments under one roof, metaphorically. Drug enforcement, licensing, and research, all of which focus on opioid drugs would certainly be able to communicate better, be aware of each other's programs, learn from each other and enable the agencies to recommend and agree on changes.

I bounced my restructuring idea off Vincent Martello, the Ulster County, NY executive I introduced readers to in an earlier chapter, and David McNamara, the fellow who manages *SAMADHI*, the treatment facility.

Vincent wasn't enthusiastic. He wrote back, "Thank you Barry. Part of the problem is that the enormous campaign financing and lobbying power of the drug companies has resulted in Congress undermining the DEA and other federal agencies and I think that simply having a czar is, at best, only part of the solution. An example of this can be found at the links below. *60 minutes* and the WP [*Washington Post*] did a joint investigation that revealed that Congress defunded the highly successful DEA unit that was responsible for going after serious opioid prescription distribution abuses and crime. The problem is systemic with the drug companies, bad actors in the provider community, med schools, distributors, pill mills, and pharmacy corporations all being part of the problem."

I think his comments, while discouraging, support my idea that a politically savvy cabinet-level leader could be effective. Furthermore, he points at pharmaceutical industry interference with DEA's efforts. I think the billions collected for rehabilitation and aftercare from the industry defangs their power base. It may even make the public pay attention more.

Vincent and David have forged a private/public partnership that is working well. I remind readers that Vincent Martello is Director of Community Relations in Ulster County, NY's Department of Health. David

McNamara is Executive Director of *SAMHDI*, a drug rehabilitation service he founded in Kingston, NY.

Vincent contracts with *SAMHDI* to provide rehabilitation for Ulster County's drug addicts and that partnership has a major impact on drug addiction in that community. I encourage readers to look at their respective websites and learn about their path-making strategy. Their goal is to provide constant aftercare for detoxed addicts.

In conclusion, drug addiction is not a societal public health problem without a solution. There are solutions, but they don't involve doing nothing because a hundred thousand people a year die from overdose and the rate is increasing. Some of the solutions are available to us, for example drug detox in an in-patient setting that cleaned up Peter, the former addict I interviewed. The key to successful detox though, is not simply to get the patients off drugs and release them. It requires aftercare, and that aftercare must be available to the public, not limited to people who can afford it.

Remember, addicts are patients, not criminals. They need therapy, not prison.

Drug smuggling must stop. Interdiction is fragmented between Customs and Border Patrol and DEA, and the result is the flow of illicit drugs into the U.S. doesn't stop.

We should reorganize the President's Cabinet to include a secretary whose mandate is to manage the entire process. That process includes opioid drug manufacture, formulation, international trade, prescription, dispensing and the other side of the business, controlling illicit drug smuggling, sale and use.

Write your congressional representative.

References

Books

Mate, Gabor
> In the Realm of Hungry Ghosts, Close Encounters with Addiction

Mann, Charles C.
> 1493, Uncovering the New World Columbus Created

Lovell, Julia
> The Opium War, Drugs, Dreams, and the Making of Modern China

Chouvy, Pierre-Arnaud
> Opium, Uncovering the Politics of the Poppy

Avorn, Jerry
> Powerful Medicines, The Benefits, Risks, and Costs of Prescription Drugs

Goldstein, Avram
> Addiction, From Biology to Drug Policy

Goodman, Louis S. & Gilman, Alfred
> The Pharmacological Basis of Therapeutics (2nd Edition)

Articles

Eisenstein, Toby K.
> The Role of Opioid Receptors in Immune System Function

Frontiers in Immunology, December 2019

Sukel, Kayt

In Sync: How Humans are Hard-Wired for Social Relationships

Dana Foundation, November 13, 2019

Stetka, Bret

Important Link between the Brain and Immune System Found

Scientific American, July 21, 2015

Centers for Disease Control and Prevention

Drug Overdose Deaths Rise, Disparities Widen

Accessed July 21, 2022

Young, Simon N.

The Neurobiology of human social behavior: an important but neglected topic

J. Psychiatry & Neuroscience, September 2008

Hutchinson, Mark R. & Watkins, Linda R.

Why is Neuroimmunopharmacology crucial for the future of addiction research

Neuropharmacology, January 2014

Butler, Stephen F. et al.

Development and Validation of the Current Opioid Misuse Measure

Pain, July 2007

Johnson, Mathew W.

Pilot Study of the 5-HT2A R Agonist Psylocybin in the Treatment of Tobacco Addiction

J. Psychopharmacology November, 2014

Nolen, Stephanie

Fentanyl From the Government? A Vancouver Experiment Aims to Stop Overdoses

NY Times, July 26, 2022

Canadian Drug Policy Coalition
Harm Reduction Saves Lives and Connects
People with Vital Social Support services and
Evidence-Based Treatments
Accessed July 26, 2022

Vancouver Coastal Health
Harm Reduction
July 27, 2022

Gordon, Elana
Lessons from Vancouver: U.S. Cities consider
supervised injection facilities
The Pulse, July 5, 2018

Feeley, Jef
Teva Pharmaceutical Will Pay Over $4 Billion
in Opioid Settlement
Bloomberg.com, July 26, 2022

CDC.gov
NCHS Data on Racial and Ethnic Disparities
Accessed March 2019

Fridell, Mats et al.
Prediction of psychiatric comorbidity on
premature death in a cohort of patients with
substance use disorders: a 42-year follow-up
BMC Psychiatry, May 15, 2019

Jerome, Lisa, Schuster, Shira and Yazar-Klosinski
Current Drug Abuse Reviews, 2013

Council of the Inspectors General on Integrity and
Efficiency
Combatting the Opioid Crisis: Role of the
Inspector General Community
October, 2019

www.ingramcontent.com/pod-product-compliance
Lightning Source LLC
Chambersburg PA
CBHW031847200326
41597CB00012B/300